LIVESTOCK SHOWMAN'S HANDBOOK

SECOND EDITION

LIVESTOCK SHOWMAN'S HANDBOOK

A GUIDE FOR RAISING ANIMALS FOR JUNIOR LIVESTOCK SHOWS

SECOND EDITION

by Roger Pond

Pine Forest Publishing

Second Edition
First Printing

Photographs by Roger and Connie Pond

Publisher's Cataloging in Publication
(Prepared by Quality Books Inc.)

Pond, Roger.
Livestock Showman's Handbook : a guide to raising animals for junior livestock shows / Roger Pond. -- 2nd ed.
p. cm.
Includes index.
Preassigned LCCN: 96-070178
ISBN 0-9617766-3-3

1. Livestock--Handbooks, manuals, etc. 2. Livestock--Showing--Handbooks, manuals, etc. 3. Livestock exhibitions--United States--Handbooks, manuals, etc. I. Title.

SF65.2.P65 1997 636.08'88
QBI96-40388

Published by Pine Forest Publishing
314 Pine Forest Road
Goldendale, Washington 98620

Printed in the United States of America

ACKNOWLEDGMENTS

It would be impossible to mention all of those who have contributed to this book, but the author would like to recognize a few who have provided considerable information and assistance.

These include the many county agents and extension specialists who are the source of much of the information in the book. Special thanks goes to Eddie Thomason for writing assistance and information about dairy cattle and Leath Andrews for his assistance with writing and compiling swine information. Thanks also to John Fouts, who has answered many questions and spent countless hours searching the files for materials the author was sure he had seen somewhere.

The author is grateful to Gwen and the late Theo Caldwell for their expert advice on all aspects of sheep production and for providing excellent photo opportunities.

He wishes to thank local veterinarians for being very generous with their knowledge and subscribing to the theory that educating clients is good business.

A special thanks is due the editors and staff of *Western Farmer-Stockman* magazines, where most of the book originally appeared in the author's *Growing & Showing* column. Information was also contributed by a number of livestock producers, including Neil and Jill Kayser and family, Jim and Sharon Pond, and others too numerous to mention.

The author is also grateful to Connie Pond for the original design of the book — and for photographing all of those animals to which she is allergic.

DISCLAIMER

Every effort has been made to make this book as complete and accurate as possible. The book is not meant to be the last word on feeding and care of animals, however.

Information in the *LIVESTOCK SHOWMAN'S HANDBOOK* is derived from university recommendations and research results wherever possible; but is not intended to give advice on veterinary procedures or treatments. Competent professionals in veterinary medicine or nutrition should be consulted if advice in these fields is desired.

The author and Pine Forest Publishing shall assume neither responsibility nor liability for losses or damages caused, or said to have been caused, by recommendations contained in this book.

TABLE OF CONTENTS

Introduction	ix
Part I – BEEF CATTLE	
Selecting Market Steer Projects	3
Feeding Market Steers	8
Breaking to Lead	13
Fitting Market Steers	17
Fitting Beef Heifers	27
Showing Beef Cattle	30
Steer Judges Can't Afford to be Human	34
Selecting Beef Heifers	37
Warts and Ringworm	41
Feeding and Management of the Yearling Beef Heifer	44
Feeding Beef Cows	48
Assisting Cows with Calving	51
Part II – SHEEP	
Why Sheep	57
Selecting Breeding Ewes	58
Feeding Breeding Ewes	62
Fitting Breeding Sheep	68
Lambing Difficulties	74
Feeding Orphan Lambs	77
Selecting Market Lambs	81
Care and Feeding of Market Lambs	85
Fitting Market Lambs	90
Showing Market Lambs	95
Wool Care	101

Part III – SWINE

Selecting Market Pigs 107
Feeding Market Pigs 111
Fitting and Showing Pigs 116
Breeding Swine Projects 121
Feeding Breeding Sows 124
Baby Pig Care 128

Part IV – DAIRY CATTLE

Selecting a Dairy Heifer Project 135
Care and Management of the Dairy Heifer 141
Feeding Dairy Heifers 145
Fitting and Training Dairy Heifers 149
Showing Dairy Heifers 153

Part V – FEEDS & RATIONS

Ruminant Nutrition 159
Hay Quality 163
Management of Small Pastures 166
Feeding Wheat 170

Part VI – JUNIOR SHOWS

Winning and Losing 179
Buying Market Project Animals 183
Reading Carcass Data 186
Hiring Livestock Judges 189
Junior Livestock Sales 192
Careers in Agriculture 195

APPENDIX
Glossary 201
Registry Associations 206
Index 210

Introduction

Parents of 4-H and FFA members often agree animals are like insanity: We get them from the kids. Then, when a person gets half-way confident with one species of livestock, someone goes out and buys a different one.

All is not lost, however. There is a wealth of information on the feeding and care of livestock, if only we knew where to find it.

That's where the *LIVESTOCK SHOWMAN'S HANDBOOK* fits in. The *LIVESTOCK SHOWMAN'S HANDBOOK* provides a quick reference for the many young people, parents, and advisors who work with junior livestock projects.

This Second Edition contains more photos and new information on fitting and showing beef cattle; as well as new chapters on Feeding Beef Cows, Ruminant Nutrition, Feeding Dairy Heifers, Baby Pig Care, Buying Livestock Projects, Hiring Livestock Judges, and Careers In Agriculture.

The book melds scientific information of value to all livestock producers with the special needs of the junior livestock member, small scale producer, and backyard sheepherder. The animals have the same basic needs regardless of how many we own, but the facilities and techniques for meeting these needs will vary.

The author's technical training and more than 15 years experience as a county extension agent and vocational agriculture teacher provide valuable insight into the problems faced by junior livestock members and their parents.

Most of the information in the *LIVESTOCK SHOWMAN'S HANDBOOK* was compiled from the author's *Growing & Showing* column, which has appeared in numerous farm magazines and publications throughout the U.S. and Canada.

BEEF CATTLE

Selecting Market Steer Projects

Market steer selection is the crucial first step in producing a winning entry for any steer show. You can mix all sorts of special rations and feed like a pro; but if your steer isn't correct in conformation and size, you'll still have that blank space in the trophy cabinet.

I should hasten to add that the chance of winning a trophy is not the most important thing youngsters receive from a steer project. Those who measure success in terms of grand champions are almost certainly headed for disappointment. The reason the grand champion is the grand champion is because the judge says he is. It's something to strive for, but let's keep it in perspective.

I'll also toss in the observation that many 4-H and FFA members achieve satisfaction from showing animals they raised themselves or that Dad produced on the farm or ranch. They may even prefer the cute one with the floppy ears. Heaven forbid!

Before heading out to buy a steer the junior livestock member should have some idea what size animal will be needed for a particular show and approximately how much this critter will cost. Club leaders and parents can help the breeder by letting him know what size and type of animals the kids are looking for, before the youngsters arrive at the farm.

If the breeder can sort and pen a number of animals, this

provides a better opportunity for selecting a project. It's hard to select animals from a large group in a feedlot or field, and it's easier to keep the kids' attention if some of the better animals have been sorted into a smaller corral or pen.

Junior livestock exhibitors should be looking for a steer with enough frame to grow to the desired weight without being over-fat, and one that is likely to grade choice at this weight. Most youth shows have a minimum weight for steers and some also have a maximum weight.

Most judges seem to be looking for a steer weighing from about 1200 to 1350 pounds.

There's always some debate about what the meat packer wants and whether shows are in tune with current markets, but most show judges seem to be looking for a steer in the 1200 to 1350 pound range. The recommended size and type of steer will vary with different parts of the country; so it's a good idea to check on the show weights of winning steers in your area the last few years.

The frame score chart shown below is one way to estimate the proper slaughter weight for a particular animal. Those with experience can make this estimate by the eyeball method, but using the chart and taking some measurements can be a good educational exercise, either before or after project animals are purchased. It doesn't hurt to recalibrate the experienced eyeball occasionally, either.

Expected Slaughter Weight		750-850	851-950	951-1050	1051-1150	1151-1250	1251-1350	1350
	Frame Size	**1**	**2**	**3**	**4**	**5**	**6**	**7**
AGE IN MONTHS	**7**	36	38	40	42	44	46	48
	8	37	39	41	43	45	47	49
	9	38	40	42	44	46	48	50
	10	39	41	43	45	47	49	51
	11	40	42	44	46	48	50	52
	12	41	43	45	47	49	51	53
	13	41.5	43.5	45.5	47.5	49.5	51.5	53.5
	14	42	44	46	48	50	52	54
	15	42.5	44.5	46.5	48.5	50.5	52.5	54.5
	16	43	45	47	49	51	53	55
	17	43.5	45.5	47.5	49.5	51.5	53.5	55.5
	18	44	46	48	50	52	54	56

The height in inches shown under each frame size is the minimum height for that frame size.

The chart is used by taking a hip height measurement in inches, with the calf standing on level ground. One simply lays a yardstick across the hips of the steer and measures from that point to the ground. (Putting the calf in a scales or chute works well.)

Applying these measurements plus the animal's age to the chart gives an estimate of the proper slaughter weight for individual steers. This predicts the weight at which a steer of a particular frame size can be expected to grade choice.

Looking at the chart you can see that a calf that is 12 months old and measures 45 inches at the hip is a frame size "3" and would be expected to finish at a slaughter weight of 950 to 1050 pounds. This is much too small for the desired show weight.

On the other end of the scale a frame size 7 may not fit your desired show weight, either; unless you decide a 1350 or 1400 pound steer is O.K. for that show. In some parts of the country a frame-size 7 may be just what you are looking for.

If we say a slaughter steer should weigh 1250 to 1350, the chart would say we are looking for about a frame size "6". Again the desired frame size for show steers varies with time and in different parts of the country, but the principle remains the same.

How old should the steer be? First he has to be old enough to meet the desired weight. This generally means a steer between 15 and 22 months of age at show time. Animals in the older part of this range are more likely to produce the necessary marbling in the ribeye muscle to meet the choice grade, but those in the lower part of the age range are also capable of grading if fed properly.

Let's say for example that the calf is 12 months of age, has a hip height of 49 inches, and weighs 800 pounds six months before your county fair. If we want him to weigh 1250

This judge has his work cut out for him.

pounds at the Fair, he has to gain about 450 pounds over the 180 days before the show.

These are realistic numbers that permit adjustments in the feeding program to achieve a desired weight and finish at show time. Of course we have ways of checking up on a steer during the feeding period. It's a good idea to weigh him once a month if possible.

If livestock scales aren't available, the steer's weight can be estimated by measuring heart girth and length as described in the "Feeding Market Steers" section.

Feeding Market Steers

There is truth to the old saying, "The eye of the master fattens the cattle". We should update our terminology, however. Spokesmen for the beef industry say consumers don't want fat on their meat, and the eye of the master must be calibrated accordingly.

Experienced steer feeders will agree that selecting the right size and type of steer is the first step in avoiding excess fat at show time. Adjusting the feeding program to achieve the desired show weight is the second step.

If you are new to the junior steer feeding game, you'll receive plenty of advice on how to feed your steer; and you may sometimes be surprised by the inconsistencies in recommendations. If we can keep an eye on the basics of steer nutrition, this advice doesn't seem so contradictory.

Many judges say they prefer steers weighing 1200 to 1300 lb. that have .3 to .4 inches of fat over the ribeye. These steers need to be trim and are expected to have enough marbling within the ribeye muscle to hang a choice carcass.

Most feeders of steers for 4-H and FFA shows divide the feeding period into two parts: A growing phase in which the steer is expected to gain at a modest rate without beginning to deposit fat; and a finishing phase during which the animal gains at a faster rate and deposits fat, both externally and within the muscle.

The length of the finishing phase is often 120 to 140 days. This will vary in different parts of the country and will change as markets change. Length of the growing phase depends upon what time of year the steer is obtained, the age of the steer, and the date of the show this critter will attend. A major goal for the growing phase is to have the steer at the proper size when the finishing period begins (about 120 days before your show date, for example).

This may be a good time to insert some disclaimers. We should recognize that some steers will need more than 120 days on a finishing ration before they will make the choice grade. Others may need less. Some exhibitors may feed the same ration for the last six months before the show.

We do have to start somewhere, however, and this is probably the most common practice for the management systems and project requirements in many areas.

A growing ration includes more hay and less grain than the finishing ration, making this feed higher in fiber and lower in energy. Many feeders also prefer a higher fiber concentrate, such as oats or beet pulp for growing rations. Cost and availability of particular grains should be a prime consideration.

A growing ration for steers should contain 14 percent protein or higher for the total ration. This often means high quality hay with a relatively small amount of grain concentrate. Younger animals need a higher protein ration than older animals.

A practical growing ration is a grain mix of 1/3 barley, 1/3 corn (or wheat, if the price is right), and 1/3 oats, fed at the rate of about four to six pounds per day, along with all of the alfalfa hay the animals will eat. A steer can be expected to gain about 1.5 pounds per day on this ration.

If grass hay or poorer quality alfalfa is fed, a protein concentrate will be needed to provide adequate protein for younger steers. Many junior livestock members may prefer to buy a prepared grain mix. This may be the most economical alternative in many situations.

Weigh the steer as often as possible during the feeding period.

If we buy a steer December 1, for a late August show, we have about 8 1/2 months to get the animal to show weight and condition. If we divide this into two phases, we would likely feed the steer on a growing ration for 4 1/2 months and gradually convert to a finishing ration for the last 4 months of the feeding period.

If our steer weighs 700 lb. and is put on a hay and grain growing-ration in mid-December, the animal can be expected to gain 1.5 lb. per day, or 180 lb. over the 120 days of this period.

This would put the steer at 880 lb. on April 15, and leaves 120 days for feeding the higher energy finishing ration. If the animal gained 3.0 lb. per day during the 120 day finishing period he would weigh 880 plus 360 = 1240 lb. about mid-August.

Of course, we have ways of checking up on a steer during the feeding period. It's a good idea to weigh him once a month, or more often if possible. If scales aren't available, the steer's weight can be estimated from the following heart girth and length measurement chart.

Heart Girth and Length Measurement for Steers

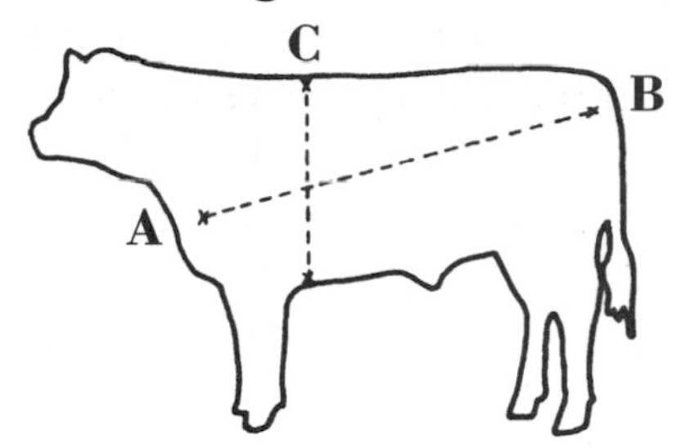

1. Measure the steer's circumference (heart girth) in inches, at a point slightly behind the withers (**C**).

2. Measure length of body from the point of the shoulder to the pin bone (**A-B**). (Note "point of shoulder" is not the top of the shoulder.)

3. Multiply heart girth X body length and divide by 300 for estimate of weight in pounds.

Heart girth X body length ÷ 300 = weight in pounds

Source: The Stockman's Handbook, M.E. Ensminger

The steer is converted to the finishing ration gradually, by decreasing the amount of hay in the ration and increasing the grain concentrate portion by 1/2 pound or less per day, until the animal is eating only three to five pounds of hay per day and is receiving all of the concentrate he will clean up.

A commercially prepared grain ration is often the most practical for junior steer exhibitors. The finishing ration should contain 12 percent protein or higher. For those who prefer to mix their own grain concentrate rather than buying a commercial mix, the following is a commonly recommended finishing ration:

Corn (or Wheat)	25%
Barley	50%
Oats	10%
Beet Pulp	10%
32% Protein Supplement	4%
Trace Mineralized Salt	1%

Steer feeders often prefer a good quality grass hay or a grass-alfalfa mixed hay for the finishing period. There may be less chance of bloat or digestive problems with the grass hay in the mix rather than straight alfalfa.

Many variations of this ration are possible, depending upon grains available and current prices. Corn and oats will be more available in many areas than wheat, barley, or beet pulp. Check your 4-H and FFA manuals and talk to experienced feeders to learn about recommended ration alternatives.

Steers will generally do better if two or more are fed together. Regular feeding times are important, and the animals should be observed for a few minutes at each feeding to be sure all are eating.

If a steer is not eating properly and goes "off feed", his ration should be cut back to half the amount normally given. The ration is then increased gradually until the animal is eating the required amount again. Fresh water and trace-mineralized salt should be available at all times.

Don't guess at the weight of your feeds. Use the scales.

Breaking to Lead

It was such a relief to see a black steer running down the road. This was spring weigh-day for 4-H steer projects, and the runaway steer told me I had found the right place.

The steer was dragging a rope halter and doing his best imitation of a stampeding buffalo. When I jumped from my car and tried to head him off, the critter made a right turn into the neighboring pasture.

The boy who owned the steer finally got close enough to grab the lead rope, and then hung on like a bulldog while the other 4-H'ers ran along behind shouting encouragement. The youngster dug his heels into the wet pasture, causing the steer to change leads, but he didn't slow down.

It had rained all morning, and the grass was slick, so the 4-H'er didn't mind being dragged around the pasture like a human anchor. It was quite a show — sort of like a scoop shovel race without the shovel. Finally the steer got tired and was subdued.

This was a small county show, with only eight steers to weigh, but the steer in the pasture was tame compared to the one that got away. I don't mean the steer that got loose; I mean he got away!

That steer was last seen going down the road in a westerly direction and was finally captured in a neighboring township a couple of months later. He showed up at the county fair with a log chain and two large men draped around his neck. He escaped once more before submitting to the indignity of being judged.

We scheduled spring weigh-days as an educational event for 4-H members, and we learned plenty. Some days a person can only stand so much education.

It would be a shame to have all these experiences without learning something; so let's take a look at what can be done to prevent those impromptu rodeos.

One of the first things I noticed when we began weighing steers several months before junior shows was that we had fewer runaways at the county fair. There were plenty of escaped steers at the weigh-days, but this is always less embarrassing than chasing them through the carnival.

The second thing I noticed is junior exhibitors who start breaking their steer early have far fewer problems than those who wait until the last minute.

Breaking a steer for a junior show begins with selection. A steer with a bad temperament can be a disastrous experience for a junior exhibitor. No matter how good the animal's conformation may be, it's not good enough if his

It's best to begin with a steer of good temperament.

temperament is bad.

Sometimes a steer doesn't show a personality problem until you pen him and try to teach him to lead. This is one reason for beginning to work with a steer early. If the animal can't be handled, it's best to get another one.

You may have noticed that the wildest steers often belong to the older exhibitors. I think this shows that it's not how big you are that's important, but how much time you spend training the steer.

Most exhibitors recommend haltering and tying the steer as the first step in breaking. Some suggest tying the steer for just a short time in the beginning, and gradually increasing this time over a period of several days. Many will leave a rope halter on the animal and allow the steer to drag the lead rope after he is turned loose. This helps acquaint the steer to the pull of the halter when the animal steps on the rope.

Putting the halter on the animal, tying him up, and taking the halter off will help gentle the steer. A strong rope halter that tightens under the animal's chin when pulled is best for this purpose. The halter should be made of material such as nylon rope that won't shrink when wet.

Strap halters that don't tighten or exert pressure are not good for either tying the animal or leading it. These are often used for tying dairy heifers to prevent hair being rubbed off by the halter; but beef steers generally need something stronger.

A strong pen and a strong halter with a long (10 to 20 foot) lead rope will prevent a lot of problems, and keep the steer from practicing his escape routine. You might begin by leading the steer to water or feed, but this should be done within a secure pen, just in case the animal isn't hungry or thirsty at the moment.

Many steers have learned to lead after being tied to the back of a tractor or truck and led around in this way. Safety precautions regarding how the animal is tied and where it can put its feet are important, of course.

One booklet suggests that tying the animal to a donkey is

the easiest way to teach a beef animal to lead. If you don't have a donkey, the other methods may have more appeal.

Some steers learn that an 80 lb. 4-H member is fun to butt around; and although they are not mean or wild, they give you fits. An equalizer in the form of a short broom stick is generally recommended. This is applied gently to the bridge of the steer's nose each time he becomes rude.

This should be used only when the steer butts and in the manner of training while leading — rather than to settle a grudge. We want the steer to associate the broom stick with his bad habits instead of with the person who is leading him.

After the steer is gentle enough to lead well, the animal can be taught to stand properly and become acquainted with the show stick. A well trained steer has a better chance of achieving a top placing, because the exhibitor is able to show the animal to its potential.

The animal should learn how to get into a truck and go for a ride sometime before the show. It helps to get the steer used to people and being hauled around, before the day you load him up and take him into the fair.

After the steer is gentle enough to lead well, the animal should become acquainted with the show stick.

Fitting Market Steers

Fitting and showing is one of the most exciting elements of 4-H and FFA steer projects. While many beginning exhibitors have older club members or parents to advise on fitting practices, there is plenty of room for panic for those who are fitting a steer for the first time.

Beginners learn fast, though, and attendance at a few steer shows will provide enough information to make the newcomer competitive.

Most exhibitors begin the steer fitting process with brushing and training the animal's hair several weeks before the show. The idea is to brush hair in the direction it doesn't want to go — until it stands out and stays where you want it.

The amount of hair your steer has during the summer months can be increased by keeping him inside during the day and turning him out at night. Some exhibitors wet their

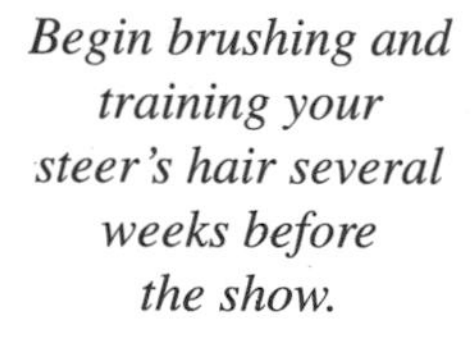

Begin brushing and training your steer's hair several weeks before the show.

cattle with water before turning them out in the evening to increase hair growth. Hair growth varies with the steer, of course. Some will grow hair and others won't.

Hooves should be trimmed two to four weeks before show. This allows sore spots to heal, if any are created.

Clipping the steer is generally performed a week to 10 days before the show. Preparation for clipping includes brushing and combing the steer. Then, you should look him over and decide what you want to accomplish with clipping. When in doubt, it's always best to have advice from an experienced exhibitor before getting too active with the clippers.

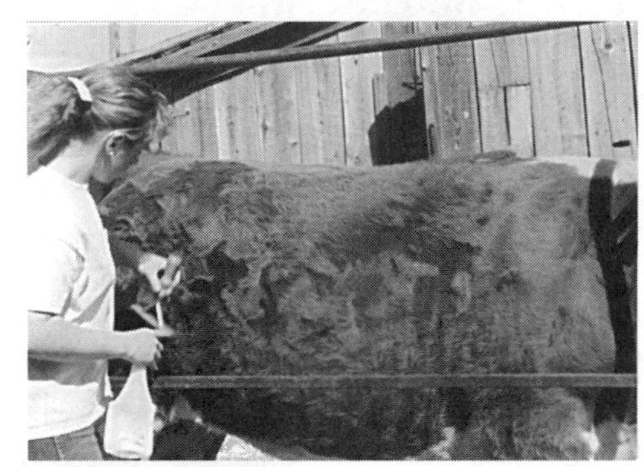

The idea is to brush the hair the direction it doesn't want to go.

Hooves should be trimmed two to four weeks before the show.

Most clip the steer's entire head, beginning at the poll. Many clip to a point several inches behind the ear (about one clipper width), and then blend this down into the lower neck and brisket. Some clip the outside of the ear, while others suggest no clipping on the ears, or thinning only. The inside of the ear should not be clipped.

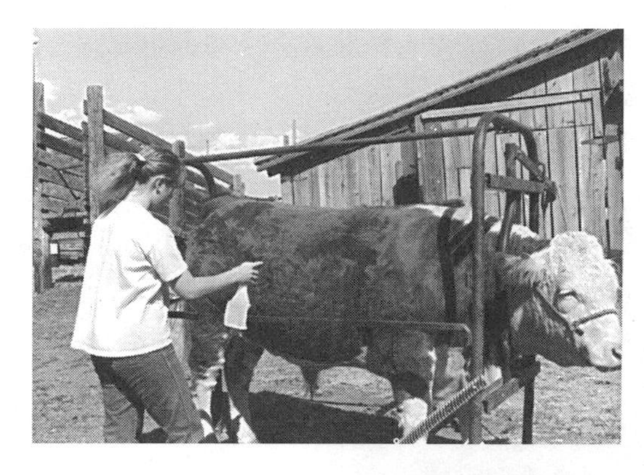

Brush and comb the steer. Then look him over and decide what you want to do with the clippers.

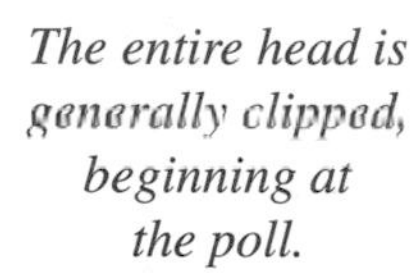

The entire head is generally clipped, beginning at the poll.

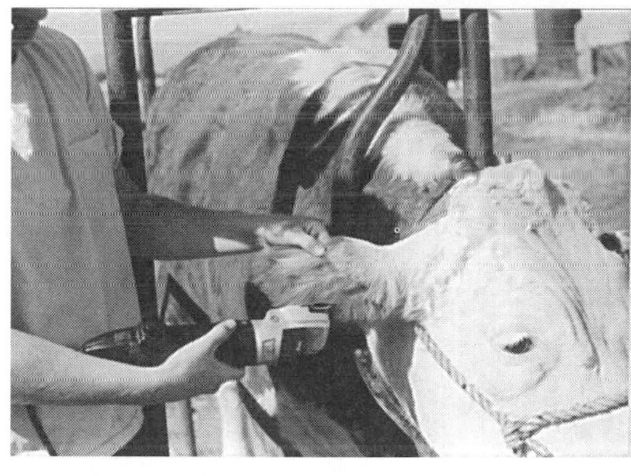

Clip the backs of the ears and long hair on the edge of the ears. Leave hair inside the ears.

The brisket and belly is generally clipped and blended in with the rest of the body. The modern steer is supposed to look trim, and clipping of these areas is designed to give that effect. Experienced exhibitors also clip areas on the shoulder, legs, and topline to give particular effects or blend areas together.

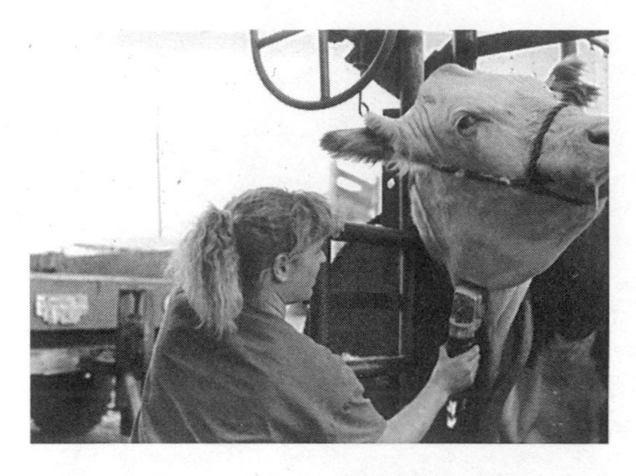

The neck and brisket is clipped and blended back into the shoulders.

Clip the entire underline on most steers.

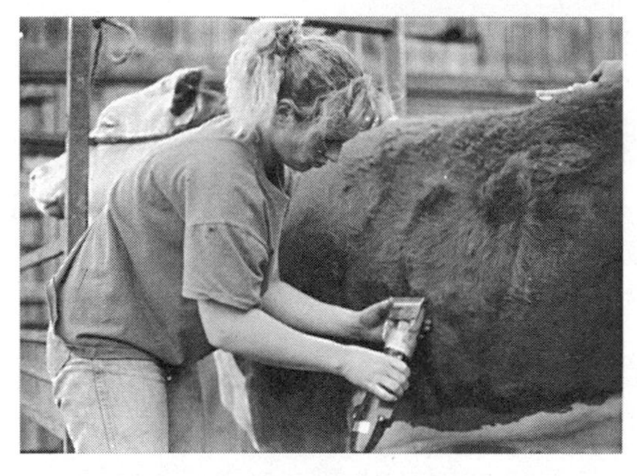

Turn the clippers over and blend closely clipped areas of the underline into the longer hair on the sides.

Styles vary for clipping of the tail and tailhead. Some clip the tail from the twist area to a point several inches below the tailhead. Others may clip only a few inches on the tail. Longer hair is left at the top of the tail and around the tailhead. This hair is clipped and combed to level out the topline and to give the appearance of body length.

The tail is generally clipped from above the twist to several inches below the tailhead.

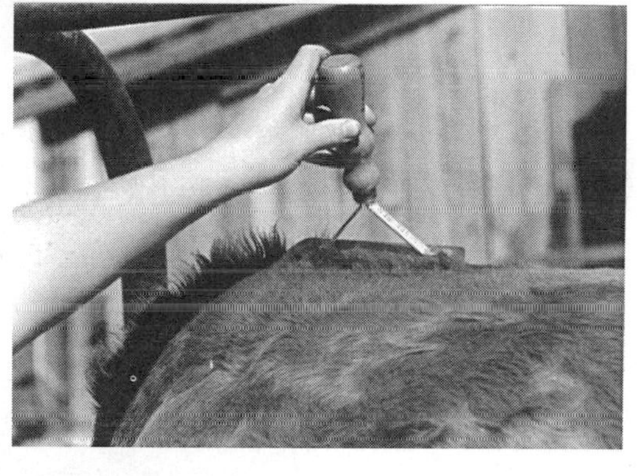

Use your scotch comb to pull the hair up on the back and tailhead.

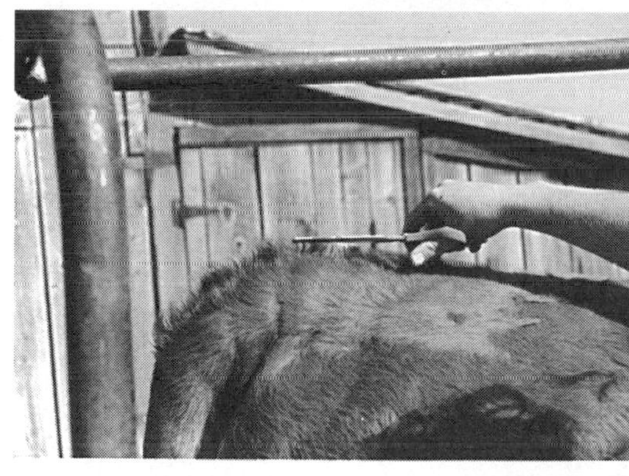

Clip high spots on the topline with scissors or clippers. Don't clip too closely at the tailhead.

Tail switches are commonly formed into a ball by combing, spraying with commercial preparations, and tying up the switch to the shaft of the tail. Some prefer to comb the tail switch into a more natural form.

These things are related to current styles, but many also have a practical value for improving the appearance of the

Begin by brushing the switch out with your scotch comb. Then, hold the end of the tail switch and brush hairs back toward the tailbone (called ratting).

Still holding the end of the switch reach into your pocket for the plastic tail tie you remembered to bring along.

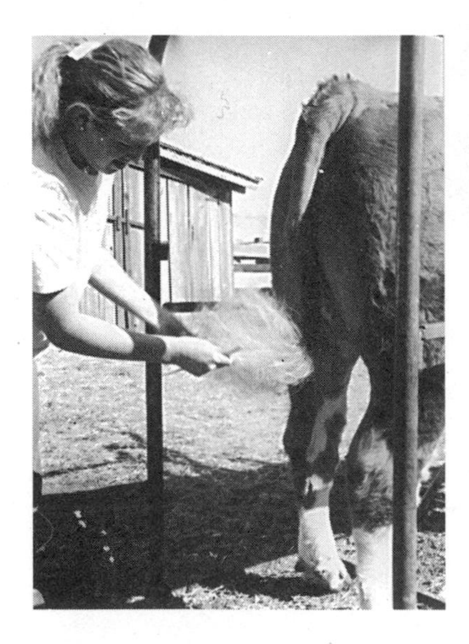

Pull the switch up – forming a ball – and insert the plastic tie through the tip of the switch.

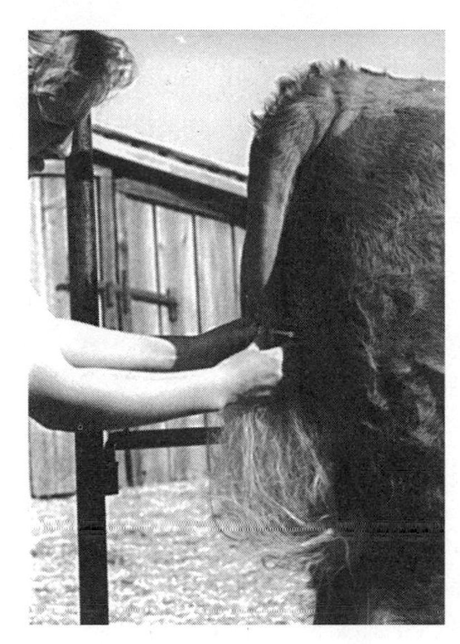

Tie the tip of the switch to the tail bone and cut off excess plastic.

Spray the switch ball with hair spray (adhesive) for a finished job.

animal. The short, puffy tail jobs help to make a steer look taller. Some say if you pull the tail switch up too much the steer will look heavy-fronted.

Everything you plan to do at the show should be practiced at home. There should be few surprises the day of the show.

When show day arrives, your steer should be fed, watered and bathed well ahead of time. You many be able to wash your steer the day before show, and then spot wash on show day.

A light feeding and limited amount of water is recommended the day of the show. How much feed and water will depend upon the steer, and the space between feeding time and show time.

The final fitting begins with tying the tail switch as practiced at home. Apply hair-spray to the switch to hold the hair in place. Then, bone-up the feet and legs with a commercial wax, saddle soap, or other product made for this purpose.

Hair wax is applied to the tailhead, and the hair is combed and trimmed as needed to add straightness, length, and height.

Many exhibitors use a hair wax, such as peach wax, showfoam, or hairset spray, on the animal's body to hold the hair after brushing. Then, they comb the hair up to make the animal look thicker or deeper where desired.

Now, you are ready for the ring. Make sure your animal is set up at all times, and be sure to watch the judge and ignore the crowd. If your mom gets nervous, give her a dollar so she can go buy a corn dog.

Bone-up the feet and legs with a commercial wax, saddle soap, or other product made for this purpose.

Apply wax to tailhead. Comb and trim hair to add straightness, length, and height.

Apply a hair wax, such as peach wax (or a showfoam; or a hairset spray) to the animal's body to told the hair up after brushing.

Comb hair up to make the animal look thicker or deeper where desired.

Watch the judge.

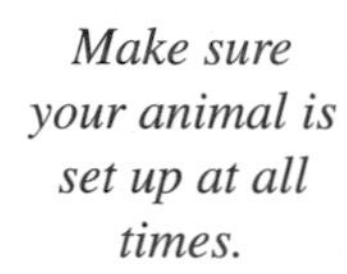

Make sure your animal is set up at all times.

And be sure your jeans have a pocket for your scotch comb – and that big, Champion ribbon you plan to win!

Fitting Beef Heifers

While market steers are the glamour project at many shows, we need to remind ourselves that breeding heifers are fun to show, too. Breeding heifer projects outnumber the steers in some areas, and may be a better project for many youngsters.

Beef heifers are smaller and easier to handle than steers, making them more desirable for younger exhibitors; and the heifer can be a long-term project, while a steer doesn't have much of a future.

Preparing a beef heifer for show is nearly the same as fitting a market steer — with a few minor differences. The first and most obvious difference is the feeding program. While steers are fed a high energy ration to achieve a certain amount of condition and fat cover, the heifer should be fed for growth and much less condition.

The heifer's ration should provide adequate protein and energy for growth and an improved hair coat, but a high energy ration can be detrimental. Excess energy consumption lowers a heifer's milking ability as a cow and her lifetime production potential.

The feeding program for a yearling heifer is likely to include good quality alfalfa hay and a few pounds per day of a relatively high-fiber concentrate such as oats or beet pulp. The amount of concentrate needed will vary greatly with individual animals, but two to six pounds of concentrate per day is a reasonable range for many yearling heifers. The vitamin and minerals content of the ration is also important.

The first job in getting the heifer ready for show is to be sure the animal is trained to lead and stand correctly. As with steers, a well-trained heifer is essential for a good job of showing. Training a heifer to lead is generally easier than training a steer (often because junior exhibitors start working with the heifer at a younger age).

Health problems, such as warts or ringworm, should be watched for and treated early. Treatment for worms is also a good idea in many areas.

An attractive hair coat is beneficial, of course. One way to increase hair growth is to keep cattle inside during the day and turn them out at night during the summer months. Many exhibitors wet cattle with water before turning them out in the evening to increase hair growth. Daily brushing improves the hair coat, also.

If the heifer's hooves need trimming, this should be done four weeks before the show to permit time for healing of any sore spots that may occur as a result of trimming. Whether hoof trimming is needed will depend upon the individual heifer and the conditions under which she is raised. Younger animals may not need their hooves trimmed for junior shows.

The pro's begin training the animal's hair and clipping a heifer several months before the show. The younger exhibitor will more likely save the clipping job for the last few weeks, when help is available from a club leader or experienced exhibitor. If an extra heifer is available, the young exhibitor can practice on the extra animal to gain added experience.

The overall goal in clipping the beef heifer is to give the appearance of trimness and femininity. This begins with clipping the belly and brisket, the neck, and the throat area.

The amount of clipping and trimming on the heifer's head depends upon the effect desired. All clipping of the head or throat area is blended into the clipped areas of the neck to give a smooth transition.

The beef heifer's head is generally not trimmed as closely as a steer's would be. A common approach in many breeds is

Beef heifers make an excellent project for younger exhibitors.

to smooth up the head rather than clip the hair short. Some trim the hair on the back of the ears and others prefer to leave this hair.

Some leave hair on the poll, to be combed into a topknot. This makes the heifer look taller, and a little bit like Woody Woodpecker. By the time you read this, the topknot fad may have faded away. I certainly hope so.

A certain amount of clipping is done on the legs, shoulders, and topline to smooth out these areas or to give various effects. Hair on the tailhead is trimmed as needed but left long enough to be combed-up, giving the appearance of greater height or length-of-body.

Tail switches may be formed into a ball by combing, spraying with commercial preparations, and tying the switch to the shaft of the tail. As this is written there is a trend in some areas toward combing the tail switch and leaving it long and natural, rather than tying it up.

Styles come and go. Check with experienced fitters, your vo-ag teacher, or 4-H leader to see what folks are doing in your area.

Showing Beef Cattle

Amid all of the excitement of clipping, brushing, and preparing a beef animal for show, it's easy to forget that showmanship ability is at least as important as having a well-fitted animal. A good job of showing and a clean, well-fitted animal are essential for any beef cattle class — not just the showmanship class.

You will need three pieces of equipment for the show ring: A show halter that fits the animal correctly, a show stick, and a Scotch comb or flat comb — to be carried in the rear pocket of your jeans. If a Scotch comb is carried, the teeth should point inward to avoid poking the steer next to you and causing a scene. All of this equipment has been practiced with at home, to assure the animal and the exhibitor know exactly how each item is used.

Proper dress for the show ring is also determined well ahead of show day. Many exhibitors avoid the last minute panic of dressing for the show by changing outfits well ahead of time, and then wearing a pair of coveralls over their showmanship clothes. This prevents getting your show clothes dirty during final preparations for the ring.

It's important to be ready to enter the ring when your class is called. This requires keeping track of previous classes and having both the animal and yourself ready.

If your class has been called and you have the opportunity to be first into the ring, go for it. Don't crowd or rush to be first, but don't hesitate if the chance is there. The first animal into the ring has an opportunity to make a good impression on the judge, and that's what showmanship is all about.

The first exhibitor into the ring should lead out at a relatively brisk pace, so all animals will show at their best. Beef cattle exhibitors always circle the ring clockwise.

Beef animals are led at a brisk walk with the lead strap held in the right hand and show stick in the left. Most show halters come with lead straps that are too long for junior exhibitors. For many youngsters, the strap should be cut off to give a total length of about four feet for the complete strap and chain.

You must keep one eye on the judge and the other on your animal while showing. The judge wants to see the animal set properly, and needs the exhibitor's attention for hand signals and other directions. When your animal is stopped, the lead strap is switched to the left hand and the show stick to your right.

Exhibitors should not watch the crowd or take directions

Beef animals are led at a brisk walk, with the lead strap in the right hand and the show stick in the left.

from others outside the ring. If parents would like to get more involved in the show, they should get their own animals.

Most judges will have the animals circle the ring once and then motion for exhibitors to stop. Each time you are instructed to stop, the animal should be set with the animal's head held high and all four feet in proper position. Anytime your animal is stopped and set-up, you should try to leave several feet between animals and keep within line.

After the animals have circled the ring a time or two, the judge will line them up and inspect them more closely. Whenever judges handle your animal, wait for them to finish and move to the next; then brush up the hair where it was touched and make sure the animal is reset properly.

Placement of the feet can be accomplished with pressure on the halter in combination with use of the show stick. Pushing back or pulling forward on the halter will cause the animal to move its feet with the change in weight distribution. For some, this is more effective than trying to move the feet with the show stick.

The steer or heifer can be kept calm by scratching its belly with the show stick. This helps to keep the back from sagging, also. (You should not appear to be sawing the animal in half, however.)

When instructed to change places in line, the exhibitor should lead out forward and turn the animal to the right (clockwise). The animal can then be led back through the spot in line just vacated and led to the proper place. Always avoid making very short turns, as this detracts from the animal's appearance.

Alternatively, the animal can be led around the end of the line if that is more convenient. Judges in many areas don't care which way the animal is turned when changing places in line. Check on local custom if there is a doubt.

The judge's first line-up is generally not the final placing. Exhibitors should not be discouraged or quit showing if they are near the bottom end of the line. You should never quit showing, no matter where you are placed!

It's important to continue showing until the judge has given reasons and the class is dismissed. I have seen many situations where no one knew for sure which end of the line was first place, until the judge began giving reasons.

When the class is dismissed, the ring steward should instruct exhibitors to lead out in line, with the first place animal leaving the ring first. This looks better and is safer for exhibitors and spectators than a disorganized stampede from the ring.

Good sportsmanship is important outside the ring as well as when showing. Criticizing the judge's placing insults those who won, and is unfair to the judge.

Showing livestock is an educational activity. We can always learn something from the judge's placings and reasons.

Keep one eye on the judge and the other on your animal while showing.

Steer Judges Can't Afford to be Human

Of course they are only human, but we can't let them off with an easy excuse like that. Anyone can be human. Only a select few can judge steers.

Years ago we had judges who could line-up a set of steers so quickly they were a marvel to watch. These folks would speak fondly of the smooth and mellow steers at the top of the line, and drive triumphantly out the gate in their Lincoln Continentals — smiling to everyone as they passed. The steers went out the back gate in a truck, and everyone was happy as could be.

But now, we have carcass data, and many of the old time judges became human, while others gave up the game to the young and courageous. Nothing is more destructive to a judge's self-confidence than a set of carcass data.

Into this dank and gloomy pit steps the modern steer judge (who drives a Chevy and leaves the motor running). The judge endeavors to place the steers in such a way that each animal in the line is slightly better than the one below him, and somewhat inferior to the one above. Judges do this by virtue of their keen eye, superior mental discipline, and undisputed possession of more guts than anyone else on the fairgrounds.

Judges never fail this test. They always get the steers lined-up, somehow; and select for the Grand Champion a clean and lovely beast, considered to be a vision of bovine perfection. Then they jump in their Chevys and head for the gate.

Let's be fair and give judges credit for lining up the steers. They could have left them milling around the arena, to be rounded-up later by the sale committee. I'm sure many have considered this option.

So now the carcass data arrives, and look what happens. The Grand Champion grades "Select" instead of "Choice," the reserve champion goes "Standard," and of 48 steers only 12 grade better than "Good." How can we explain this sad state of affairs?

I'm not sure I can explain it, but I'm going to try. Not being a steer judge myself, I'll only attempt to describe what the judge is up against (and hope I don't offend anyone in the process).

Let's start with why only 25 to 50 percent of the steers in many junior shows will grade Choice, while feedlots expect much higher percentages. Some would say the answer is obvious: Show steers are fed differently, hassled all summer, hauled to the fair and mixed with strange cattle, washed, led around the show ring, and thoroughly humiliated by somebody with a microphone. Then, they are hauled to slaughter with a strange bunch of pampered and pestered pets.

A steer judge sees a lot of rear-ends during the show season.

Many of the things we do when exhibiting cattle cause stress. This stress can affect the amount of fat deposition (marbling) within the meat, and thus cause the cattle to receive a lower carcass quality grade.

Judges try to estimate the fat covering over the back and ribs of a steer, as well as deposits of fat in other parts of the body, as indicators of finish and marbling. The judge uses these estimates for predicting carcass grade.

Other factors affect marbling, as well; including heredity, age and maturity of the animal, and various aspects of the feeding program. A judge looks for cattle that would gain well and be economical to raise, as well as producing a desirable carcass.

Toss in the fact that a trim "Select" or "Good" grade carcass is often more valuable to the retailer than an over-fat "Choice" carcass, and we can see the judge has been asked to do more than just assure the Grand Champion grade choice.

I think most judges look for a steer to have at least three-tenths of an inch of fat cover over the ribeye to assure a reasonable chance of grading choice. Beyond this, there is always a certain amount of risk that the Champion will not grade.

There is no evidence that placing a Grand Champion blanket on the back of a steer will destroy marbling in the ribeye muscle, but one has to wonder sometimes.

Selecting Beef Heifers

A good start is important in any enterprise. This is especially true when selecting breeding cattle. An error in selection of market animals corrects itself within a few months, but mistakes made when purchasing breeding cattle often create long-term problems.

Youngsters growing up on a ranch or farm where there is a good herd of cattle are doubly blessed when it comes to selecting animals: They have a ready source of good stock and family members who know what it takes to produce profitable cattle. This chapter is written for those who are not blessed with a lot of cattle raising experience, and whose main interest is selecting beef heifers for 4-H and FFA projects.

One of the most important decisions is which breed to buy. Although there is no best breed of beef cattle, there are advantages to having animals that are in demand in your area. Choosing a popular breed assures more opportunities for sale of offspring and a better selection for replacement animals.

I don't mean to be critical of new and exotic breeds. A number of imported breeds have proven themselves through performance and become well-established over many years. On the other hand, we should recognize that along with these great successes, we've seen a lot of "wonder cattle" that didn't make the grade. The trick with having a popular breed is to have one that is still popular when you have some to sell.

Once a decision about breeds has been made, a reputable cattle breeder can give the prospective buyer good advice as well as a selection of quality cattle from which to choose. A reputable breeder can be located through advertisements in beef magazines as well as talking with others in the cattle business.

While conformation is still important, especially for a show heifer, today's cattle breeders also have very good production records and are happy to explain what this data means. When selecting heifers, the index of the dam and performance records of the sire will be the most important information.

If buying a purebred animal, the junior exhibitor should not expect to buy the best animal in the herd. We should remember that performance indexes only compare animals within that herd; and a good breeder will have many good animals. We should also expect to pay a premium when selecting animals from a group. If the breeder is selling one of his better animals, he needs a better than average price.

Selection is made easier if the breeder will confine the animals in a corral or enclosure where you can see them easily and have plenty of room to watch them walk. We want correctness in the feet and legs and a heifer showing good femininity. Heifers with short heads; short, thick necks; and very heavy muscling lack femininity and are not good breeding prospects.

There is a trend toward putting more emphasis upon femininity, milking ability, and maternal traits in beef cattle. This is not new for many breeders, but represents a swing away from the heavy muscled, "bigger is better", philosophy of some years ago.

We want a heifer with good size for her age without being over-fat. Research has shown that yearling weight is one of the best indicators of a beef heifer's lifetime production potential. On the other hand, heifers that have been fed a high energy ration when young will not produce as well over their lifetime as those which have been fed a less-fattening ration. We want to avoid over-conditioned animals.

Calving-ease is a very important trait in breeding heif-

A nice hereford heifer with good size for her age.

ers, but a hard one to evaluate visually. Pelvic measurement is the best predictor of calving ease.

The health program of the herd from which the heifer is selected can be very important. A disease transmitted to other cattle at home can be costly in lost production if not actual loss of the animals. Certain diseases require extra precautions when moving cattle. Recommendations for brucellosis vaccinations should be discussed with your veterinarian to assure compliance.

Whether the 4-H or FFA member should buy a registered heifer depends upon the long-term plans for the project. Buying a registered heifer offers the advantage of knowing more about the animal's ancestry, which increases the predictability of her performance. The pride of owning a registered animal that might be the start of a purebred herd has value, too.

A major disadvantage of registered stock for 4-H and FFA members may be the initial cost and the difficulty of recovering this higher cost through a later sale of the animal or her offspring. Sometimes the breeder from whom the heifer is purchased will assist in finding a market for registered animals produced. (This shouldn't be expected, however, unless this person volunteers to help.)

In addition to lower initial cost, crossbred heifers offer higher fertility levels (on the average) and advantages in many other production traits. 4-H and FFA members wishing to show a crossbred heifer should check local show regulations, however.

Some county fairs and junior shows require all breeding stock to be registered, as do breed shows. County shows may permit entry of animals that meet breed standards, even if these animals are not registered. Some shows have classes for crossbred cows and heifers, also.

Warts and Ringworm

Warts are a common affliction of beef cattle. While they aren't damaging in most instances, warts can lead to infection through open wounds and reduce the value of hides as a result of the animal's rubbing. If warts are very large or in a sensitive location, they may contribute to other animal health problems.

Besides, warts are against the rules for beef cattle exhibited at most county fairs and junior shows around the country. That means we have to get rid of them, one way or another.

Warts are caused by a virus that is transmitted by direct contact with infected animals or indirectly through contact with contaminated feed bunks, stalls, needles, eartag applicators, etc. The virus is considered highly host-specific and there is very little danger of transmission between different animal species. Humans will not get warts from cattle or other animals (despite what we may have heard about toads).

The two most common treatments for warts in cattle are vaccination and surgical removal. Vaccination takes about 90 days to remove the warts, and many vets say the success rate leaves something to be desired. Two doses of vaccine should be administered 3 to 5 weeks apart.

Warts that are discovered within 90 days of the show are probably going to require surgical removal. "Surgical removal" is veterinary language for cutting them off. This is done with a sterilized instrument, such as a knife or surgical scissors, or tying them off with thread. Warts can also be scraped loose with a curry comb.

There will be some blood with these methods of removal. After warts are removed the affected area should be treated with iodine.

The vaccine has the advantage of being bloodless, but surgical removal can be performed much closer to show date. Your veterinarian should be consulted to see what he or she recommends and to be sure things are done correctly.

Ringworm is also a frequent problem in cattle and is prohibited for animals going to shows. Ringworm is a fungal disease causing a loss of hair and scaly appearance of the skin. Younger animals are most susceptible, but cattle of any age may contract the disease.

Although the ringworm organism common in cattle can also infect humans, the danger of transmission from cattle to humans is not very great.

Animals affected by ringworm develop lesions about three weeks after infection. The infection progresses into scaly patches from a half-inch diameter to several inches across. Infections are most often found around the eyes, ears, neck, and tailhead. A loss of hair is caused from the animal's rubbing as well as damage by the organism.

Because it occurs more frequently under damp conditions and in confined animals, ringworm is more common in wetter climates than in drier regions. The problem is more frequently seen in the winter than during pasture seasons.

Because ringworm spores can live several years in barns and stalls, one of the best preventions is keeping animals out of those places when possible. Sunlight is considered the best treatment in many areas, but winter sunlight is not reliable in most of the country.

The ringworm fungus is also spread by direct contact and may be transferred from one animal to another on combs, brushes, blankets or halters. Lice may move the fungus to other animals as well as transporting the spores to new sites on the infected animal.

Ringworm is often associated with other animal health problems, or nutrition. Animals that are doing poorly seem to be more susceptible than healthier individuals in the herd. A deficiency of Vitamin A is sometimes suspected, as it is with other skin diseases. Proper nutrition and a good parasite control program may help in prevention.

There is no effective vaccine for ringworm. Treatment generally consists of brushing the infected area with tamed iodine on a toothbrush. My veterinarian tells me "tamed iodine" formulations are those without alcohol, whereas tincture of iodine contains alcohol.

Some books recommend softening crusts in the area of ringworm infection by brushing with soapy water before applying the iodine. Fungal ointments are available for treatment of this disease, and various injections are suggested in some of the older literature.

Talk with your veterinarian before trying anything. Some of the older remedies may be more entertaining than effective, and your vet can suggest treatments that have proven effective in your area.

Feeding and Management of the Yearling Beef Heifer

The period between one and two years of age is critical for the proper growth and development of a beef heifer. The nutritional needs of this age heifer are not especially high, but either overfeeding or underfeeding can have a detrimental effect on proper growth and lifetime production.

A good feeding program for the yearling heifer should keep her growing well through puberty and breeding age and maintain this growth through calving. At calving age we want a well grown heifer that can produce and suckle a healthy calf. We also want the heifer in adequate condition to ovulate and breed back in order to have her second calf one year after the first.

It's difficult to talk about age and weight for breeding heifers because of the variability in breeds, but we have to begin somewhere. Let's say you have a heifer of an English breed, such as Hereford or Angus, and this heifer weighs about 650 pounds at 12 months of age. Larger breeds and many animals of the English breeds will exceed this weight, but we can use this example.

A yearling heifer such as this needs to continue growing at least one pound per day to breeding age of 13 to 15 months. We would prefer she grow more like 1.2 to 1.5 pounds per day. This would put her at 700 to 750 pounds and growing well at breeding age.

Again, larger breeds and larger type heifers will exceed this weight at breeding age. While the yearling heifer needs good growth and size, we don't want her fat, however.

To obtain 1.2 to 1.5 pounds of gain per day the heifer will need good pasture. A small amount of grain (two to four pounds per day) and protein supplement will be needed if the animal is on poorer quality pasture or other forage.

If good quality alfalfa hay is fed, three or four pounds of barley or two to three pounds of corn per day could be added to the ration to achieve the desired gain. If a grass hay or poorer quality alfalfa is fed, a protein supplement will also be needed.

If the heifer was of good size at breeding, she should continue growing at a rate of about 0.5 pounds per day during the first six months of pregnancy and should gain about one pound per day during the last three months of pregnancy. During this last trimester of gestation the fetus will comprise about 0.7 pound per day of the weight gained by the heifer.

Thus, our heifer weighing 750 pounds at breeding age of 14 months weighs about 950 to 1,000 pounds at calving. At first calving the heifer should be about 80 to 85 percent of her expected mature weight.

Those numbers surely are easy to write down and juggle around, but obviously we don't have this animal gaining 1.5 pounds per day the day before breeding, 0.5 pound per day the following day, and one pound per day six months after that. Cattle are not that exacting, but these numbers help explain the general feeding pattern we have in mind.

A practical summary of these scientific estimates would say the heifer should be on a good growing ration from weaning to breeding, but we don't want her fat. She can stand a lower plane of nutrition after breeding but should continue to grow.

The last three or four months of pregnancy, the heifer needs a better feeding program because the fetus is developing rapidly. A good feeding program at this time also assures that the heifer will be in good condition at calving and

Good quality pasture is an adequate ration for yearling heifers and beef cows nursing calves.

will be more likely to cycle and breed back for the next production cycle.

There is a long-standing controversy among cattlemen as to whether heifers fed too well during late gestation will give birth to larger calves and, therefore, have more calving problems. Although most experts would agree that overfeeding during gestation is a mistake, research has shown that inadequate nutrition during this period can lead to weaker heifers, which may have more trouble calving and may not cycle for rebreeding as they should.

The calves may be weaker, also, with less chance for survival. Trying to reduce calf size by providing inadequate nutrition to the heifer is generally a losing proposition.

Bull selection and breeding can be a special problem for junior livestock producers or anyone else with a small number of animals. Artificial insemination can solve some problems for the small producer but requires good management and ability to detect heat as well as good technician service. The advantages of artificial insemination include the quality of bulls available and a reduction in the spread of disease which can result from moving animals from farm to farm.

The other option is taking the heifer to the bull or bringing the bull to the heifer. Either way, it's important to breed the heifer to a bull that will sire calves which are relatively small at birth. The owner of the bull may have to advise on the question of birth weight.

We should also be concerned about the health program of the farm or ranch providing the bull, as well as the vaccination status of the heifer.

Your veterinarian should be consulted about recommended vaccination programs for your area. Many states require brucellosis vaccination for all breeding stock, but regulations vary by state and are prone to change on short notice. Your vet can advise on current requirements in your state.

Feeding Beef Cows

Printed information on beef cattle production falls into two disproportionate piles. We can get advice for wintering a herd of cows on 200 acres of corn stalks, or we can read how to support the suburban cow on pigweeds pulled from the garden.

Somewhere in between we find the hobby farmer or FFA member with a few beef cows, a small pasture, and little time for pulling up pigweeds. The small herd owner has to decide whether to identify with the commercial farmer with a big herd or the agricultural survivalist with a bin full of veggies.

I recommend the small herd owner take the information designed for the commercial farmer, as you may have guessed. For my mind, it's easier to scale down the rations of the bigger producer than to separate the pigweeds from all of the other things cows don't like very well. Either way, the owner of a cow or two needs to know the nutrient contents of his feeds, as well as the nutritional needs of that particular cow.

Let's first consider the needs of the cow according to her age, size, and production status. A young beef cow with her first calf has the highest nutritional needs. This first-calf heifer should be about 85% of her expected adult weight when her calf is born, as described in the previous chapter.

This means she has to continue growing after calving, as well as feed the calf and maintain her own body. We also want the young cow to ovulate and breed back within the first three months after calving. Research has shown that body condition and level of feeding before and after calving are major factors in how soon after calving first-calf heifers will cycle and breed back.

Because she is still growing, the first-calf heifer needs a ration with a higher protein level than the ration we might feed older cows. Adequate protein is important for milk production as well as growth and maintenance. The young cow also needs a ration with a higher level of energy.

This is where having a small number of cows is both an advantage and a disadvantage. The small producer generally has the advantage of having the cows well-confined and can improve the ration by feeding grains for more energy, or protein supplements if needed. On the other hand, a person with only two or three cows finds it impractical to pasture and feed young cows and old cows separately.

A typical ration for a first-calf beef heifer weighing 950 lb. just before calving would be about 22 to 26 lb. of good quality hay, containing 12% protein or more, per day. (I know hay containing 12% protein is not "good quality" to some people, but it is better than hays often fed to beef cattle.) Three or four pounds of corn or barley could also be fed each day, if needed to increase the energy level — or if the heifer won't consume the quantity of forage needed.

Good quality pasture can substitute for the hay in the ration above. Late summer pasture, which may be low in protein, or pasture without adequate growth should be supplemented with hay or better quality forage of some kind.

The heifer's requirements will increase after calving, and higher quality hay and possibly some grain may be needed. The best sources of advice on these matters are your local extension agent, vo-ag teacher, or experienced livestock producers. Local conditions and types of feed available are always important in designing rations.

Mature cows have much lower requirements for maintenance during gestation than the pregnant heifer. The nutritional requirements of the mature cow during early gestation can generally be met with poorer quality roughage.

These cows will likely be larger, however, and their total nutrient needs will increase in late gestation and after calving to about the same level as those of the two-year-old with her first calf. The mature cow can utilize poorer quality forage, but she needs more of it.

The protein requirement of the mature cow during early lactation is about twice the level needed during gestation. Her energy requirement is about 50% greater after calving than it was during gestation.

The mature cow's greater rumen capacity gives her an advantage over the heifer in ability to consume and digest higher quantities of roughage. She is more likely than the heifer to meet her requirements without the need for supplemental grains or higher quality roughage.

Minerals are also important to good nutrition and may be provided through trace mineralized salt and other mineral supplements. The level and balance of minerals needed varies with areas of the country as well as the grains and forages fed. The amount of calcium and phosphorus in the diet and the relationship between these two minerals can be a concern under some conditions. Grasses and hays are generally good sources of calcium, while grains are high in phosphorus.

The old habit of throwing out a block of salt and expecting this to take care of mineral needs is not recommended in most areas. Nutritionists and veterinarians generally recommend cattle be fed loose salt and minerals, rather than blocks, to achieve the consumption needed.

Vitamin A deficiency can be a problem if green forage is not available, or if hay more than one year old is being fed. Again, a good source of information is the local veterinarian, who knows the signs of this deficiency and understands what can be expected under local feed conditions.

Assisting Cows with Calving

Researchers tell us over 70 percent of calf losses are associated with calving difficulty. While the person with only a few cows may not assist with calving often, he or she will often have to decide whether assistance is needed. For the small-time operator, deciding if calving assistance is needed may be as important as knowing how to help.

My personal experience with cows and calving is limited to an FFA project many years ago. Therefore, I won't pretend to be an expert on the subject. I do have the distinction of never owning a cow that would calve unassisted. She had two, and the vet pulled them both.

The size of calf in relationship to the size of the cow or heifer is the single most important factor in calving problems. Although pelvic measurements of heifers are getting increased attention for prediction of calving ease, these techniques are not always practical for the owner of a few cows. The person who plans to buy a heifer or two would do well to select from a herd where calving ease is an important consideration in the breeding program.

Let's remember that it takes two to tango when it comes to calving difficulties. The heifer or cow is one part of the problem and the bull is the second component.

In addition to the size and development of the breeding heifer, calving problems are greatly influenced by the bull that sired the calf. A beginner would do well to avoid the calving difficulty that can result from breeding to the biggest bulls, without also considering the calf birth weights expected from these bulls.

We can't just blame the breed of the bull, either. Calf size from individual bulls varies greatly within breeds.

Even though a high percentage of calving problems may be preventable, most folks who raise cattle will have to give assistance sooner or later. In order to know when to assist or when to call the veterinarian for professional help, the owner needs to understand what is normal. A good reference on this subject is the book, *Cow-Calf Management Guide — Cattleman's Library*, available from cooperative extension offices and universities.

The normal birth process in cattle is generally described as three stages. The first stage begins with a general uneasiness and signs of slight pain, such as the cow kicking at her belly. Cows on pasture will seek an isolated place.

During the latter part of this time, uterine contractions become stronger; and straining is evident. The first stage is so variable in length I hesitate to say what's normal. *Cattleman's Library* says, "This stage lasts for an average of two to three hours in a cow and four to five hours in a heifer, although it can continue normally for longer periods."

The second stage of calving is characterized by more serious straining, and the animal becomes almost oblivious to her surroundings. A lot of things happen during this period, including positioning the calf for birth, entrance of the fetus into the dilated birth canal, rupture of the water bag, and expulsion of the calf through the vulva.

This sounds simple enough, but quite a few things can go wrong during the second stage. Luckily, most of them don't.

There are several dangers of assisting the cow too early, including injury to the cow or to the calf. If things seem close to normal, the natural process is best.

The third stage is expulsion of the afterbirth. The afterbirth is normally expelled within a few hours, but may be retained longer. Retention beyond 24 hours is usually con-

A healthy calf should be up and suckling within an hour or so after birth.

sidered abnormal and may require treatment.

Some people say there are four stages to calving — the fourth stage being, "Call the vet!" Whether a person should give assistance or call for help varies with the owner's experience and equipment.

I asked one of our local veterinarians for a personal opinion on this question. He suggests the cow should certainly deliver within four hours after the water bag is presented. If not, she needs some help. The water bag will usually be broken before it is observed.

If delivery progresses to the point that a part of the calf is showing (foot, nose, whatever), but there is no progress for an hour or more, it's time to assist. The vet also suggests that if the cow goes through three or four hours of vigorous labor without results, she needs assistance.

One of the most common mistakes seen by veterinarians is trying to help too soon, before the needed dilation of the cervix and relaxation of the vagina and vulva have occurred. A second mistake is applying traction and trying to pull, even though all parts of the calf aren't properly presented. A third mistake is waiting too long (24 to 48 hours) before deciding the cow won't do it on her own.

Above all, don't panic! Terrible things have been done while trying to get a calf out of a cow. More force is generally not the answer. As one vet said, "That calf has been in there nine months already, a few more minutes isn't going to hurt anything."

SHEEP

Why Sheep?

A farm just doesn't seem complete without a few animals. For many small farms a few sheep provide just enough noise and manure to make the place interesting.

The problem with sheep is they multiply so fast the owner is under constant pressure to decide which ones to keep and which to sell. The cattle owner has two years to decide what to do with a calf, but the shepherd will be up to his haunches in woolies if he doesn't think faster than that.

Despite the mental strain, the sheep's small size and low demand for space fit well into what many junior livestock members and small-scale farmers have available. Many commercial flocks were begun with 4-H and FFA projects parents inherited when their kids left the farm.

Unfortunately, the owner of a small flock of sheep faces the same problems as those who have a few sows or a few cows. Things like how can you isolate newly purchased animals when you only have one pasture? How do you manage older ewes and ewe lambs separately if you have only one of each?

While these situations may sound extreme, similar problems are faced by many. The author still has a six-year supply of enzootic abortion vaccine, and some coccidiosis medication left over from the kids' 4-H projects. (If anyone wants to make a trade, let me know what you have.)

Selecting Breeding Ewes

Selection of breeding stock is one of the most critical decisions 4-H and FFA members have to make. It's important to look at our sheep production goals before heading out to buy ewes.

If the goal is to raise market lambs for show, we obviously want the conformation and muscling of the meat breeds. If we hope to produce wool for home spinners, there are a number of breeds with the wool quality desired.

In most parts of the country Suffolks, Hampshires, and Dorsets are the leaders in market lamb shows. These three breeds, and a few others, are the obvious choice for youngsters who want to raise market lambs for show. Excellent market lambs are also produced from crossbred ewes of the meat breeds and ewes of the medium-wool breeds, mated to meat type rams.

Junior livestock exhibitors should not feel compelled to buy purebred or registered ewes. While registered rams are the core of good breeding programs, registered ewes are not necessary and sometimes not desirable for 4-H and FFA members. Again it depends upon one's goals for raising sheep and whether the junior member has the determination and financial ability to raise and market purebred breeding stock.

There are advantages to crossbreeding for commercial production. Crossbred ewes are more fertile and more prolific, and crossbred lambs produced from these ewes have a higher survival rate than straightbred ewes and lambs.

Production records are the first requirement for replacement ewe selection. When buying ewe lambs we would like to know the animal's birthdate, weaning weight, type of birth (twin, single, triplet), and her dam's lifetime production if possible.

Lifetime production records of the dam and any available progeny records for the sire are the best indicators of the performance we can expect from an individual animal. In the case of registered stock, bloodlines, sale records, and show winnings also play a part in selection.

Production records can help tell us if this stylish Columbia ewe has the desired potential.

The old recommendation of keeping ewe lambs from sets of twins is still valid, but the laws of genetics tell us the lifetime production of the dam is more important than whether a particular lamb was born a twin. It is better to keep a single lamb from a six-year-old ewe that has averaged two lambs per year, than a twin from a ewe of the same age with a lower production average.

Ewe lambs should be selected from ewes that have produced well for their age. This means the lamb born a twin or triplet from a ewe lamb or a yearling is more likely to have the genetics for multiple births than a twin born from an older ewe, in her peak production years. On the other hand, we can tell more about a ewe with five years of records than about the ewe in her first production season.

I emphasize multiple births because research and producer experience tells us this is the single most important factor for profitability for small flock owners. A more meaningful production figure is pounds of lamb produced from the ewe within a period of time (such as 90 days), but this figure won't amount to much if the ewe doesn't have twins or triplets.

Selection for high reproductive rate is especially important in farm flocks where good feed conditions make a high level of production possible. With a high quality ration, some ewes will raise triplets by themselves, and improved techniques for feeding commercial milk replacers have made raising orphans more practical.

Many commercial sheep producers will tell you they hate triplets, and they have good reason. A bunch of bummers can be a problem in larger operations. It's hard to get a high percentage of twins without getting some triplets, however.

The small flock owner can make sets of twins from triplets. No one has figured out how to do this with a single lamb.

When purchasing breeding stock from others, the buyer must remember the health program is of prime importance. All new breeding stock should be maintained separately from the rest of the flock for observation for three to four weeks. This is sometimes difficult, but can save considerable grief and expense by preventing the introduction of disease to other stock on the farm.

Other factors to be considered in ewe selection include

Assistance from a knowledgeable sheep producer is very helpful when selecting breeding stock.

freedom from inherited defects, correct conformation, and the general health of the animal. A good reference for beginning, as well as experienced sheep producers, is the Sheepman's Handbook, published by S.I.D. Inc., 200 Clayton St., Denver, CO 80206. This publication describes the many defects common to sheep and provides helpful information on all aspects of sheep production.

Feeding Breeding Ewes

The feeding program is often the biggest problem in raising small numbers of sheep. The small flock owner often can't separate young ewes and ewe lambs from older ewes to feed them the way we would like; and when sheep are penned close to the house, there is a tendency to feed according to noise rather than condition. While the large producer is trying to economize with the feed he has available, the very small producer often has too much feed — but at the wrong time.

The major objective for a breeding ewe ration is to allow for changes in nutrient requirements of the ewe according to her age and stage of production. Let's begin with feeding of ewe lambs with the goal of breeding them to lamb at one year of age.

There are situations where feed resources or lambing dates prevent breeding of ewe lambs, but most small flocks have adequate feed available to grow lambs to breeding size and condition to lamb their first year. This requires that the ewe lamb have adequate growth to ovulate and breed at seven to eight months of age.

It was once common practice for many purebred breeders or junior livestock members who show breeding stock to delay breeding ewe lambs until their second year, allowing them to attain more growth as yearlings. This is a decision for the breeder, based upon his situation. In the case of 4-H and FFA members we must recognize that delaying breeding of ewe lambs sacrifices one year's production and postpones the question of how productive this ewe may be.

Unless junior exhibitors are on the purebred show circuit and have special need for delaying breeding, they should be breeding ewes to have their first lambs at about one year of age. Some of the more prolific ewes will have twins at one year of age.

We want a well grown lamb at breeding age, but we should not feed the replacement ewe lamb as we would a market lamb. In cattle it has been shown that creep feeding of young heifers reduces their lifetime production as a cow. It seems likely that ewe lambs can also be fed to grow too rapidly, reducing their later production potential. I haven't seen research to confirm this, but I expect many experienced sheep producers would agree that excess feeding of replacement ewes is a bad practice.

What is overfeeding? The best example I can think of would be when we feed ewe lambs for a market livestock show, but then decide to take them home for breeding stock instead. Productivity of these ewes may have been reduced by overfeeding.

Well grown ewe lambs should generally be bred their first season. A high percentage will lamb at one year of age.

Nutritive requirements of the ewe vary greatly throughout the year. These requirements are determined by stage of gestation, lactation, or maintenance needs. If we assume a late summer or fall breeding season, we can outline the ewe's feed requirements from before breeding to early lactation.

It is commonly recommended that ewes be put on good quality feed three to four weeks before breeding, and be gaining weight just before and during the breeding season. Studies have shown that increasing the ewe's condition at this time (called flushing) causes an increase in ovulation and results in a larger lamb crop.

Some studies suggest flushing is of no value when ewes are bred in October or November. This practice is most beneficial for August and September breeding or for late breeding seasons, such as January.

While it is important for the ewe to be in good condition at breeding, most authorities suggest it doesn't take much increase in energy to achieve this in a farm flock. In many cases the ewes are already fat. Worming may be enough to achieve a flushing effect in ewes that are already in good condition. In the corn belt, sheep producers commonly feed 0.25 to 0.5 pound of corn or other grain to ewes, beginning about two weeks before breeding season for flushing.

University specialists generally recommend ewes be put on a fresh pasture, such as orchardgrass, prior to breeding and that legume pastures be avoided at this time. Legumes have been found to contain compounds which can alter hormone production and affect ovulation. Although certain species of clover are the most likely to cause breeding problems, many recommend that alfalfa pasture be avoided at this time, also.

Researchers say alfalfa can cause problems as a breeding pasture under a particular set of growing conditions, or when other factors may affect growth of the plant. Alfalfa is commonly used as a flushing and breeding pasture in the western states, partly because that's the best thing available at that time of year. (The possibility of bloat from alfalfa pasture must be considered, also.)

Nutritionists in many parts of the country say clovers, especially Ladino or white clover can be a serious problem in breeding pastures. Your best sources of advice on such matters are specialists and researchers at your state university who are familiar with local forages and conditions.

Ewes need little more than a maintenance ration during the first 3-1/2 months after breeding, as there is very little fetal development during this period. Such a ration can be provided by average quality hay of 12 to 13 percent protein, at the rate of four to six pounds per ewe per day — depending upon the size and condition of the ewe. Ewe lambs and young ewes that are still growing require a higher quality ration than older ewes.

In many small flocks the most critical time in the feeding program is the last three months of gestation. The first part of this period should be dedicated to not getting the ewes too fat, and the last six weeks to giving them enough energy and protein for increased growth of the fetus while maintaining body condition.

Most small flock owners flush ewes by feeding grain before breeding. The value of this depends upon condition of the ewe and time of breeding.

If the ewe was in good condition at breeding, we would like her to gain only a few pounds during the first three or four months of gestation. Then, we would like the ewe to gain 25 or

30 pounds or more the last 4 to 6 weeks before lambing.

It isn't that easy to turn a ewe on or off when it comes to weight gains, of course; and ration changes have to be made gradually. This is the general scenario, however.

About 70 percent of fetal growth occurs during the last six weeks of gestation, making this period crucial to ewe nutrition. The ration should provide the energy and protein needed to lamb normally and be in good condition for the early lactation period.

Insufficient energy during the late stages of pregnancy can lead to pregnancy toxemia or ketosis, which usually results in death of the ewe. Ewes carrying twins or triplets are more susceptible to ketosis due to their higher energy requirements.

Ewes bearing twins and triplets will need 75 to 80 percent more energy during the last six weeks before lambing than they did during early gestation. Ewes bearing a single lamb require about 50 percent more energy during this period.

It is commonly recommended that the ewe receive high quality alfalfa hay (or a similar high quality forage) and 1/2 to 1 pound of grain per day to increase the energy level of the ration during the last four to six weeks before lambing. With large ewes that are not already fat, and with poorer quality forage, some producers may feed higher quantities of grain.

This is variable with the condition of the ewe as well as the quality of the forage. Ewe lambs and young ewes that are still growing require a higher quality ration than older ewes. Ewe lambs are losing incisor teeth at 12 to 15 months, often coinciding with lambing or early lactation. This increases the need for a high quality ration for very young ewes.

Trace mineralized salt that includes the maximum allowable level of selenium is generally recommended for areas where this element is deficient. Regulations and recommendations for selenium feeding and selenium injections are constantly changing. Your local veterinarian should be consulted about the need for supplemental selenium in your area and the best way to provide it.

A common problem in very small flocks is having ewes too fat before the last six weeks of gestation. The question then becomes, "Do we give them grain anyway?"

The experts I talk with generally say if you have a fat ewe going into late gestation, you have to keep her fat until she lambs. As the fetus grows the ewe's energy needs increase whether she's fat or not. If she doesn't maintain her body condition during late gestation, and has to draw on fat reserves, the chances of pregnancy toxemia increase.

Common sense tells us we don't want to overfeed a ewe that is already fat, but animal health specialists generally agree the chances of pregnancy toxemia will be reduced if she gets 1/2 to 1 lb. of grain during the final month before lambing.

We should recognize there is a big difference between grains such as corn, barley, and oats in energy content. Corn furnishes the most energy of these three grains, oats the least, and barley is intermediate between the two. Oats is higher in fiber and protein but lower in energy, and may be safer to feed where overfeeding is a concern.

Let's remember all ewes aren't the same size. Some judgment is needed in deciding how much grain to feed.

This might be a good place to comment that older references are sometimes out of date on nutrient requirements for the type of animals many people are raising. As breeding and management practices have changed, we expect much better performance than we did 20 years ago. It's always good practice to check with the most recent references to be sure we're not using out of date recommendations.

Ewes must be put on the grain ration gradually and ration changes should be made slowly. All feeds should be weighed rather than measured. One quart of shelled corn weighs nearly twice as much as a quart of oats. A quart of rolled barley weighs about 40% less than a quart of whole barley.

It is generally recommended the ewe receive very little grain the first few days after lambing. The grain ration is then increased gradually to the desired amount for the early lactation period.

Fitting Breeding Sheep

Market lambs are a popular 4-H and FFA project. They are small, easy to handle, and most kids just naturally like to work with the woolly, little things. Good lambs originate from good breeding stock, though, and junior exhibitors will learn more about the sheep business if they can show their breeding stock, too.

Several years ago I received a note from a 4-H leader calling my attention to the lack of printed information on fitting and showing breeding sheep. As a result, I picked up some advice from some experienced sheep exhibitors and have included it here.

We are fortunate that livestock people are generous with information and encouragement. These people are great about sharing their knowledge and ideas.

Let's take a basic approach to fitting breeding sheep, with the recognition that learning is a continuous process. The beginner will learn a lot from his own experience and by watching experienced exhibitors.

We'll start with Suffolks because that's the most popular breed of sheep in the U.S., as well as the most common at shows. The Suffolk is termed a meat breed; so wool is not a major factor in judging, as it is in the wool breeds. Black fibers are usually considered undesirable, however.

Suffolks are shown in much shorter fleece than are the wool breeds, and fitting efforts concentrate on making the animal look tall, long, trim, and muscular, as would be desirable in a meat animal.

Many Suffolk exhibitors shear ewes and rams in the show string about 60 days before a show. This provides a fleece at show time that is easy to work with and gives the animal the desired trim and muscular appearance. The belly and legs are then shorn again a few days to a week before show. The resulting product should look long, tall, trim, and muscular (if you can get them to do all of these things at once).

A more modern trend is to shear only the neck, shoulders, brisket, thighs, and belly 40 to 60 days before show; leaving long wool on the hind legs and topline. Then, closer to show day, these exhibitors use blocking shears to blend the long wool on the back and hind legs into the shorter wool of the sides, neck, belly, etc.

The idea is to add visual depth by leaving more fleece on the topline, while creating a tall and trim look with much shorter wool on the belly, legs, etc. Longer wool on the hind legs and rump is trimmed and blended to give a muscular appearance.

Many Suffolk exhibitors use small hair clippers on the lower legs and the head to trim up these areas. Any long hair or wool is clipped from the head to give a clean, neat appearance. Hair can be clipped on the back of the ears but should not be clipped inside the ears.

One of the best ways to determine how to fit a ewe or ram is by attending breed shows or sales. Another good way to learn is to look at pictures in breed magazines or sheep publications containing sale advertisements and show pictures. Find an animal of your breed that looks the way you would like yours to look, and copy that style.

A smooth shearing job is crucial to a good fitting job. If the shearing job leaves ridges of wool, the exhibitor will have a difficult time with final trimming. When the shearing job isn't smooth, it may be necessary to go over the sheep a second time to get rid of the ridges. The best time for shearing will vary with individual sheep and experience of the exhibitor, but 60 days is a fair average.

Suffolks are washed before show and can be washed right up to show day, as long as they have time to dry and be properly fitted before entering the ring. The wool should always be carded-out after washing, and should always be wetted for carding. A small spray bottle works fine for wetting the wool.

Some folks use sheep dip (if it is still available) in water for spraying the fleece when carding. Some use a water and Borax solution, while others use plain old water. I like simple solutions, such as water.

Blanketing after washing and carding is necessary as show day approaches. This helps firm up the fleece, as well as keeping the sheep clean.

Hooves should be trimmed early (about shearing time) and kept trimmed as needed until show day. Hooves should be very clean for the show, but they do not need shoe polish. (Shoe polish is not legal in many junior shows.)

Everything said so far applies to fitting all of the meat breeds, such as Hampshires, Dorsets, Montadales, etc. There may be some minor exceptions, but a look in a breed magazine or conversation with an experienced exhibitor should clear these up.

Now, what do you do to fit the wool breeds? First, we should recognize that all white-faced sheep are not wool breeds. Second, there are three categories of wool breeds: long, medium, and fine. Third, we won't do to a wool breed what we did to those Suffolks. If we do, no one will speak to us.

In the medium wool breeds you would like to have 1 1/2 to 2 inches of fleece at show time. This means the animals are shorn at least four months before show. You don't want 12 months' fleece on these sheep, partly because it would be very hard to work with, and partly because more than 12 months of fleece is technically not legal in many shows. Therefore, we would normally shear these breeds four to six months before show.

In the medium wool breeds, the fleece is trimmed and shaped to give the desired firmness and quality.

Examples of medium wool breeds are Columbia, Corriedale, Panama, and Targhee. Natural colored (black) sheep are generally fitted similar to the medium wools.

In the fine wools and the long wools, you want a very long fleece. I suggest you write the breed association for information on these. Fine wool breeds are Rambouillet and Merino. Long wools are Lincoln, Romney, and Leicester.

For ewe lambs of the medium wool breeds shearing several months before show or an early trimming with hand shears would help get some of the "baby wool" off and make the fleece easier to work with. Columbia breeders have sometimes shorn the sheep's belly, but this can vary according to current customs. The judge may wish to see how far the wool quality carries down on the fleece.

Wool breeds are not completely washed (as this is written) because the crimp and lanolin in the fleece should be maintained. Fitting practices change, however. Check with experienced exhibitors to learn what is currently acceptable.

All foreign matter, sheep keds, and other impurities should be out of the fleece. The fleece surface should have a

When properly carded, the fleece of the dual purpose breeds should be free of vegetation and still show the desired crimp and quality.

clean appearance of uniform color, although it won't be white. When parted, the fleece should be clean and show the desired crimp and quality. Legs and small dirty areas, such as under the legs, can be washed.

Medium wool breeds are sprayed, carded, trimmed, and blanketed to give the fleece the density, firmness, and quality desired. Trimming is done with hand shears, or electric shears with a blocking comb, following the same principles of conformation as in the meat breeds. The difference is that the quality and amount of wool remaining is an important judging factor.

As mentioned earlier, a very long fleece is desired in the fine and long wool breeds. In the long wool breeds, the fleece is brushed, rather than carded. In the fine wools, a light carding of the tips of the fleece may be possible.

I don't plan to go into much detail on these breeds, as I have probably already told you more than I know. Write the breed associations; that's what they're for. A list of breed associations and their addresses is contained in the appendix.

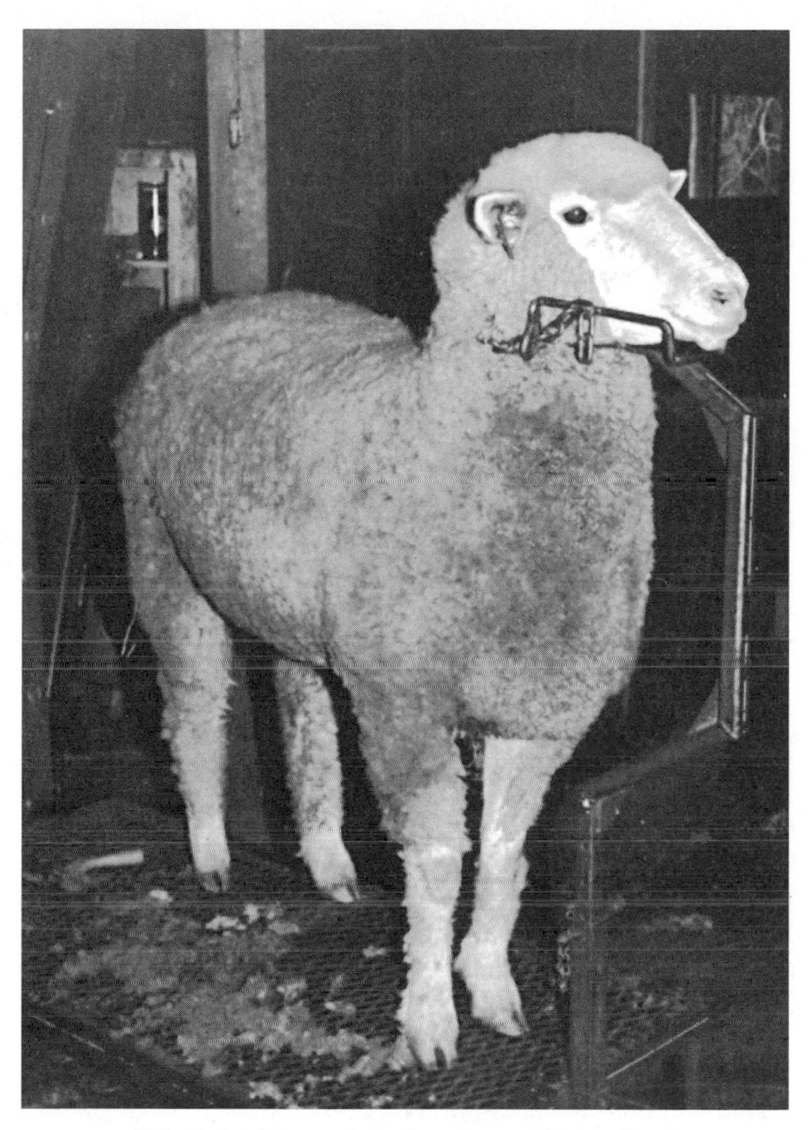

This Columbia ewe is in the process of being fitted. The head has been carded and will be trimmed and blended to give a neat, stylish appearance.

Lambing Difficulties

If all went well at breeding, we can look forward to a high percentage of twins (and possibly triplets) at lambing time. How the ewes and lambs are handled at lambing is often the key to success for sheep producers.

The most difficult lambing problem for beginners is determining what is normal lambing. This decision is further complicated by the difficulty in finding normal sheep. There are guidelines, however, that can help greatly in understanding what is going on.

Most land grant universities have printed information on the care of sheep before lambing, at lambing, and after lambing. Some years ago The University of Idaho printed an excellent series of leaflets on these subjects. Originally titled Current Information Series No. 619, these leaflets describe management practices at lambing as well as other times of the year. I would expect similar materials are distributed through county agents in other states, also.

The Idaho leaflet says, "The total time span for normal delivery is about five hours, including four hours for dilation of the cervix and one hour for delivery of the lamb. After a ewe is in hard labor for five to fifteen minutes, the front feet and nose of the lamb should appear."

Let's further consider for us beginners that these times will vary with individual animals and may be longer in ewes lambing for the first time. The pamphlet says normal birth should occur within one-half hour after hard labor starts and the water bag is ruptured.

Signs of lambing difficulty include:

A. Continued straining without appearance of a water bag.

B. Continued straining for an hour after rupture of the first water bag but without the appearance of the lamb at the vulva.

C. Partial expulsion of the lamb with the ewe unable to complete delivery.

If things aren't progressing normally, an internal examination of the ewe may be necessary. Whether you do this yourself or call the veterinarian for advice or assistance, depends on your level of experience.

Trying to help too soon is a mistake, because of the time needed for dilation of the cervix as mentioned earlier. Unassisted delivery is always best unless a problem arises.

The ideal conclusion: An attentive mother with a nice set of twins.

On the other hand, small flock owners shouldn't be surprised if they have to examine some ewes. There will be times when the sheep owner isn't sure what to do, and an examination is often the only way to find out what is going on.

Cleanliness is of major importance when making internal examinations. The person making the exam should scrub his or her hands and arms and lubricate with soap or an obstetrical cream. Soap and lubricant are important, even if throw-away plastic gloves are used. The area around the ewe's vulva should also be washed with soap.

Cleanliness, good lubrication, and gentleness are of prime importance when assisting with lambing. As someone said, "Those lambs have been in there five months already; there's no big rush in getting them out now."

When a hand (with or without plastic gloves) is inserted into the vagina, the danger of uterine infection is greatly increased. Most veterinarians recommend insertion of uterine boluses and injections of antibiotics anytime a ewe is examined internally. Specific recommendations and dosages should be obtained from your vet.

I won't attempt to give advice on correcting abnormal presentations, other than to suggest producers obtain one of the pieces of printed material available from extension offices and universities.

Sheep producers in many states can get some lambing experience in a hurry by attending one of the lambing schools sponsored by land grant universities and producer organizations. These programs provide the chance to get some expert advice as well as hands-on education.

Feeding Orphan Lambs

Sheep producers heaved a sigh of relief when quality milk replacers and cold-milk feeding systems became popular many years ago. This allowed a person to rig up a cold milk feeding system, put the orphans in a pen, and let them get their milk a little at a time — just as the ewe would do it. No more climbing out of bed at 3:00 A.M. to feed the bummer lambs.

Who would have suspected milk replacers would become so expensive, and lamb prices would get so low that sheep producers sometimes awaken at 3:00 A.M., wondering how they are going to afford the milk replacer those little devils are drinking — a little bit at a time? I don't know what the answer is, but it surely makes a person appreciate the ewe that has two or three lambs, claims them all, and produces enough milk to keep them growing.

There will always be orphans, however, even though producers put their best efforts into grafting most of these lambs to ewes that can feed them. Artificial milk replacers and cold milk feeding still offer the best chance of saving orphan lambs and turning them into something close to a profit.

Regardless of the feeding system used, it is important to give the lamb a good start by making sure it has received colostrum two or three times during the first 12 hours after birth. Timing is important because the lamb's stomach begins losing its ability to transfer the needed antibodies soon after birth and has lost most of this ability after 12 hours.

Many shepherds keep a supply of frozen colostrum milk on hand for instances when colostrum milk is not available from the ewe. Cow colostrum has proven useful for lambs, also. Antibodies can be destroyed by heating, so colostrum should be thawed slowly.

University research has shown a microwave can be used for thawing colostrum. The recommendations I have seen suggest setting the microwave at 60% power and stirring colostrum as it thaws.

Lambs should have access to a very palatable creep feed and high quality hay at a few days of age.

While the cost of milk replacers goes up and down, these products are almost always expensive compared to the price of lambs. Getting lambs on dry feed as soon as possible should be the primary goal in artificial rearing.

These lambs should have fresh water, a very palatable creep feed, and high quality alfalfa hay at a few days of age. Converting the lamb to dry feed reduces the amount of milk replacer needed, permits early weaning, and results in a healthier lamb.

Manufacturers of artificial milk replacers have excellent printed information on how to use their products. These pamphlets should be available at stores where milk replacers are sold. Even the person with only one or two orphan lambs can rig up a very simple cold-milk feeding system if this is preferred.

Lambs can be weaned as early as 4 or 5 weeks or about 25 lb. of weight if they are eating adequate amounts of dry feed at that time. When very young lambs are weaned from the milk replacer diet, a creep ration of 20 to 24% protein should be available for the first two weeks.

Two weeks after weaning the protein percentage of the creep ration can be dropped to 18%. Feeding lambs beyond this period is similar to feeding naturally reared lambs of the same weight.

Some research has shown artificially reared lambs are more susceptible to internal parasites than lambs raised on ewes. It is recommended these lambs be reared separately from naturally raised lambs if possible. A local veterinarian or experienced sheep producer should be consulted about recommended vaccinations and other health concerns for artificially reared lambs.

In addition to the health problems sometimes encountered with orphans, these lambs have a tendency to develop more belly than we would like. Cold-milk feeding systems combined with early weaning should help reduce this problem.

Sometimes a ewe will claim all of her lambs, but has more than she can feed. While larger producers may take these lambs from the ewe and feed them as orphans, those with a small number of ewes may leave triplets or quads on the ewe. Lambs that are not getting enough milk can then be supplemented with artificial milk replacer from a bottle. If the time is available, this is much less expensive than taking lambs from the ewe and feeding them as orphans.

This can be done by training one or two lambs to drink from a bottle within a few days of birth and then continuing

Two lambs from a set of quads appreciate a little extra milk from the bottle.

to feed them some milk replacer to supplement what they receive from the ewe. I am told it's difficult to train older lambs to drink from a bottle, so this training should be done within a few days of birth if possible.

My son helped one ewe raise quadruplets this way and another to raise triplets. The two quadruplets fed supplementally on a bottle consumed 25 pounds of milk replacer and were nearly the same size at weaning as the other two quads which got all of their milk from the ewe. These four lambs had a combined weight of 240 pounds at 60 days of age.

The 25 pounds of milk replacer consumed by the lambs was well worth the expense in this instance. The same ewe raised a good set of triplets on her own as a two year old.

Selecting Market Lambs

Ask four people what size and age lambs should be for a junior livestock show and you will likely receive four different answers. One of the four may say, "It all depends."

It all depends upon how you plan to feed the lamb between the time of purchase and the show date; and it all depends upon the lamb's ability to gain weight and reach the desired finish.

A well-bred market lamb, raised as a twin, creep fed, given a good health program and a good ration, can reach 120 pounds at four months of age. On the other hand, the same lamb could be much smaller if it hasn't had all of these benefits.

I believe most 4-H and FFA members are much safer selecting lambs that will be five months of age at show time. This is especially true if you plan to hold them back a little near the end of the feeding period.

Growthy lambs of six-months or older will need a lot of dieting to keep them within proper show weight. While lambs can be held at a relatively low gain for weeks if necessary, you really don't learn much about feeding lambs if they are kept on a reduced diet for weeks at a time.

A quality market lamb can be expected to gain from .5 to 1 pound per day for a 60 to 90 day period after weaning, with good management and a high energy ration. On a high quality pelleted ration, .6 to .8 pound per day gain is likely. If the lamb is on average pasture or a low energy ration, well under .5 pound per day is much more likely. (I'm speaking of weaned lambs here.)

This says a lamb weighing 80 lbs. 60 days before the show should be 110 to 130 pounds on show day. The lamb's gain will slow down a bit as it reaches its proper slaughter weight.

Most judges are looking for a properly finished lamb weighing about 110 to 130 pounds.

Because of the excitement and stress from washing, carding, and exercising, many lambs won't gain anything the last week or two before the show. Especially for summer shows, we should look at the 60 days before show as a 50 day feeding period.

When in doubt, it's best to start with an animal a few pounds heavier than what we might need under ideal conditions. It's always easier to hold a lamb back a little than to push one that isn't up to size.

Sometimes it's hard to get the lamb to a scales before

buying it, but this has to be done. If a lamb is 90 days old and weighs only 60 lbs when you buy it, we can't expect that lamb to begin gaining 1 lb. per day for the next 60 days.

The preferred weight for show lambs will vary with lamb markets in various parts of the country and preferences of individual judges. Given a good selection I think most show judges today would like a lamb that finishes from 110 to 130 pounds.

There is always some debate about what the market really wants, but we need to recognize that a livestock buyer and a livestock judge have to look at things differently. A buyer deals in large numbers and has to buy on averages to a certain extent, whereas a show judge is looking for an animal approaching the ideal.

Weigh the lamb as often as possible to keep track of its progress. Often lambs won't gain much, and may even lose a few pounds, the last week or two before show .

If the judge selects a 130 pound lamb for grand champion, he isn't saying he would like to see all lambs in the show weighing 130 pounds. He's saying, for this particular lamb, this is a good weight, and the market would be happy with a bunch of lambs like this one.

In addition to being the right size, a show lamb needs the length, scale, and muscling show judges always talk about. The best way to learn the type of lamb judges are looking for is to observe a few shows and listen to the reasons for placings. You will find that judges aren't all looking for the same thing, but that's why we show animals: To learn.

Care and Feeding of Market Lambs

The recipe for feeding lambs is simple: Build a pen, insert lambs, add feed and water, and stir until done. Like most recipes, this one leaves out the details.

Just when you think you've got all the answers, the lambs will prove you wrong. It's like a friend said several years ago, "It's amazing how fast a person can become an expert in this business. The same people who were asking me questions last year, are giving me advice this year."

The best feeding program for lambs depends upon the age and size of lambs as well as feeds available. Lambs for early summer shows may need little confined feeding to be ready on show day.

A common feeding program for late summer, fall, and winter shows is to feed a dry ration for about 60 days before show, with some variance depending upon the size and condition of the lamb at the beginning of the feeding period.

Most lambs can be expected to gain from .5 to 1 lb. per day on a quality ration. With a pelleted ration and good management a majority of junior show lambs will average about .7 to .8 lb. per day for a 60 day feeding period.

A lamb will eat about four percent of its body weight each day. An 80 pound lamb will eat about three pounds per day — a 100-pounder approximately four pounds per day. The best way to keep track of the lamb's progress is to weigh the feed at each feeding and weigh the lamb as often as possible.

Weighing the feed is extremely important when feeding show lambs. The commercial producer can feed a 100-pound bale of hay to 100 lambs with confidence the animals are getting approximately 1 lb. each, but the young exhibitor who is feeding very small numbers must weigh the feed carefully.

The most popular lamb ration in many areas is a complete pellet containing 12 to 14 percent crude protein. This ration can be self-fed, meaning feed can be kept before lambs at all times. Some prefer to hand-feed the ration twice a day.

Hand-feeding twice a day helps assure that the feed is fresh and gives more opportunity to observe the lambs in case one is not eating properly. A ration with 12 or 13 percent crude protein is often suggested for older lambs, but 14 percent protein is a better level for fast-growing lambs.

Self-feeding a complete pelleted ration works well for many lamb exhibitors.

Grain overload and bloat are frequent problems with lambs fed for show. In addition to the tendency of certain feeds to cause bloat, the problem in show lambs is often caused by feeding practices. Irregular feeding times, failure to weigh the feeds, or a lack of observation can all lead to bloat. Ration changes should be introduced gradually, over a two-week period.

Watching the lambs eat can be almost as important as feeding them. A lamb may be sick or off feed. One lamb may be hogging the feed, while others in the pen are not getting their share. The feeder may be too small, too high from the floor, or full of moldy feed.

If we have two lambs in a pen and one doesn't eat for some reason, the second lamb may get a double ration that day. Watch for scours or other evidence of digestive problems.

When grain and hay are fed, or hay is fed with a pelleted ration, it's important to feed only the intended amount. Otherwise the lamb fills up on hay and won't eat the grain. With most lambs it's best to feed the concentrate first, then the hay after the concentrate is cleaned up.

Pelleted rations can be fed as the only feed or with a small quantity of long hay. Some feeders prefer a small amount of hay (1 pound per day or less) be fed with the pelleted ration to aid with rumen function.

Grain and hay rations are also practical for market lambs, and may be most economical in situations where pelleted rations are not available or much more expensive than local

Be sure to weigh grains carefully. Rations must be formulated by weight rather than volume measure.

grain. Lambs digest whole grains very well, making processing unnecessary for grains to be hand-fed.

High grain rations are normally fed with good quality alfalfa hay, but an additional protein supplement may also be needed if a low protein grain, such as corn, exceeds 50 percent of the ration. When optimum gains are desired, the grain portion of the ration will exceed 50 percent, but should not exceed 75 percent.

Fresh water is important. Shade is critical in the summer. Trace mineralized salt will help provide needed minerals, although a commercial ration should also contain these elements.

The health program for show lambs should be a combined package with management and nutrition. When lambs are purchased from others it's important to learn the feeding and management program these lambs have been on, as well as the vaccinations and treatments they have been given.

Someone knowledgeable about sheep production should be consulted about recommended health practices for your area. This person may be your veterinarian, state extension sheep specialist, an experienced producer, county agent, vo-ag teacher, or 4-H leader.

Wether lambs should be checked carefully to be sure castration has been performed correctly. If there is a possibility the lamb's scrotum contains a retained testicle or a hernia, don't buy the lamb. If either of these a problems are discovered later in the feeding period, your veterinarian should be consulted.

Extremely short tail docks can create problems, also. Ultra-short tail docking increases the risk of rectal prolapse and should be avoided when purchasing or raising feeder lambs. Short docking is one of those show ring fads that goes against common sense.

Enterotoxemia is the number one killer of feeder lambs. Lambs should be vaccinated for enterotoxemia C & D at the time of purchase or before putting them on feed. Some lambs may need two treatments if these vaccinations haven't been given earlier, or if the timing of earlier vaccinations is open to question. Talk with your veterinarian to confirm timing for these vaccinations.

Coccidiosis is a widespread problem in lambs. Lambs infected with coccidiosis will grow poorly, and this disease can cause death in some cases. Most feed dealers have, or can get, lamb feeds that are medicated for prevention of coccidiosis.

Soremouth is a concern throughout the country and can be prevented with vaccination. A lamb that may not exhibit the disease at the time of purchase may be a carrier and develop symptoms after taken home.

This disease is very contagious and spreads rapidly through contact with infected sheep. Soremouth can be transmitted to humans, also.

Labels and directions for all medications should be read carefully to be sure animals are not treated too close to slaughter. If lambs need treatment for external parasites, this must also be done with the right products and according to label directions. Prevention of illegal residues in lamb meat is extremely important to the sheep industry.

It's always best to feed at least two lambs together. They like company.

Fitting Market Lambs

Emphasis upon leaner, more muscular animals has revolutionized fitting and showing techniques for market lambs. We've come a long way since the days when blocking meant just what it says.

Junior shows are still adjusting to new techniques and attitudes about fitting lambs. While some shows are debating how much wool lambs need for a number "1" pelt, others are getting out the clippers and shearing them to the skin.

In many cases junior lamb exhibitors are caught in the middle, with literature that tells them how we did it a few years ago, and club leaders who are doing their best to find out how we're going to do it this year. Rapidly changing techniques and lagging literature-revision can lead to moments of panic for those who do as the book says — and later learn no one else does it that way.

Regardless of how lambs are fitted, junior exhibitors should understand there's more to preparing a lamb for show than just carding, clipping, and washing. Even when lambs are freshly shorn near show day, there's still a lot of training and fitting needed to produce a winning entry.

When slick shearing first became popular for junior shows, many parents and leaders thought this would reduce the fitting effort and skill required of junior exhibitors. It does, of course; but there's still a lot of training and fitting needed to be done. The big problem in many areas is having access to a good shearer, so all of the kids have a reasonable chance to compete.

One advantage of slick shearing is that the judge has a much better view of the lambs. Another is the fact that Grandpa and Aunt Bessie don't have the chance to fit the kids' lambs for them. A disadvantage of slick-shearing is the loss of a No. 1 pelt. A No. 1 pelt was worth only about $2 the last time I checked, so this isn't a great problem.

Many shows slick-shear lambs very near show day.

Despite the trend toward shearing project lambs very near show day, junior shows in many areas still exhibit lambs with a certain amount of fleece on them. Find out your show requirements before doing anything.

If your show doesn't require or permit slick shearing, lambs are generally shorn 30 to 60 days before show. This is a good practice even if it's not required. A young lamb that has never been shorn is very difficult to do a good trimming job on; and a lamb with 60 days of fleece has enough wool to permit considerable carding and trimming.

This lamb was shorn 30 days before show, and then fitted with hand trimmers.

Whether lambs are shorn 60 days before show or very near show day, a smooth shearing job is important. This saves a lot of extra work in the final trimming and fitting. (Long wool and medium wool breeding sheep are another subject.)

Most exhibitors will shear the lamb's belly within 10 days of the show if the animal hasn't been slick-shorn. Many also shear the legs of market lambs to a point above the hock. Hand shears are used to blend shorn areas into the rest of the body.

Some shows require a minimum length of fleece for market lambs, and this must be considered when trimming. Otherwise, the most important thing is that the fleece is short enough that the lamb feels firm when handled.

Advice of experienced lamb fitters is helpful here. Pictures in breed magazines or other publications will also give an idea of the effect we are trying to create.

Junior exhibitors should allow plenty of time for washing, carding, and trimming. Younger 4-H members will tire quickly when they have to card and work the hand shears for a few hours. Twenty or more hours of washing, carding, and

A good shearing job is important, whether the lamb is shorn 60 days before show or slick-shorn very close to show day.

trimming time is not out of reason for lambs that have 60 days of fleece.

When carding and trimming a dry lamb, it helps to wet the fleece with a spray bottle containing water — or a rag dipped in water. Some people use a small amount of borax in the spray water. Others prefer to add 1/4 teaspoon of sheep dip (if it's available) to the spray mixture. I wouldn't buy a can of dip for this purpose alone. Just plain water works pretty well.

Nearly everyone washes market lambs as often and as close to the show as needed. This should begin about two to three weeks before the show. Lambs that are not slick-shorn near show time will often need two to three washings. Special attention should be given to areas where dirt collects, such as under the legs.

Dry the lamb with towels after each washing and then card and trim the wool while the lamb is still damp. Carding while the fleece is damp helps to straighten wool fibers and results in a firmer, smoother trimming job.

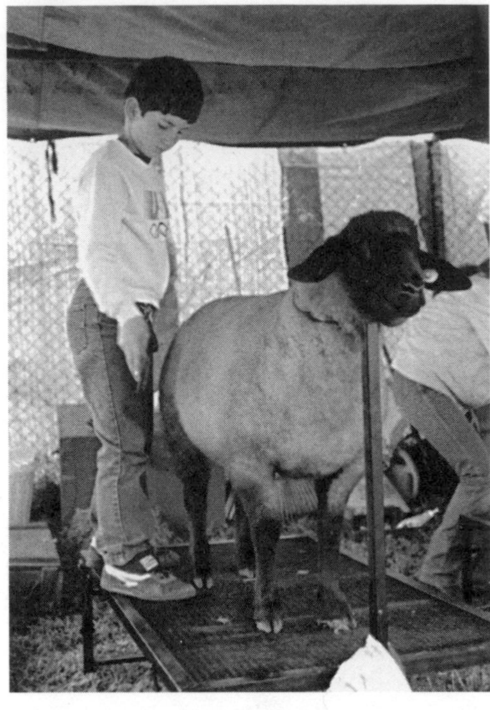

A trimming stand puts the lamb where you can work on it and keeps it under control.

Blanketing the last week or two before the show also helps keep lambs clean and firms up the fleece. Keep the lamb in a clean, well-bedded pen after washing.

Fitting stands are helpful for all types of washing and fitting jobs. Once you get the animal up on the stand, it's pretty much immobilized and will become used to being handled and worked with.

They may jump off a few times, and you shouldn't go too far away when the lamb is on the stand. (A lamb can break its neck or strangle if it jumps off the trimming stand.) The lamb can be tied to a wooden fence or post for washing, carding, and trimming, but a trimming stand makes things a lot easier.

The lamb's feet should be trimmed during the feeding period as needed. The feet may need to be trimmed twice during the month before the show. Ideally, the last hoof trimming should be a week before the show to permit time for healing of any mistakes or sore spots.

Showing Market Lambs

Preparing a lamb for show is more than carding, clipping, and washing. There are a number of other essential tasks, including having the animal tame enough to handle and in proper show condition.

The previous chapter describes the equipment and techniques for fitting your lamb. Now, let's talk about training and final preparations for showing. This takes some equipment too, but it doesn't need to be elaborate.

A trimming stand is very helpful for taming a lamb as well as for fitting. Once you get the animal up on the stand, it's pretty much immobilized and will become used to being handled and worked with.

Some exhibitors like to lift the lamb's feet while it is on the stand or tied securely. This helps to tame the animal and accustoms the lamb to having its feet moved, as they will be in the show ring.

Nearly all junior lamb exhibitors use halters for training. These can be purchased, but they can also be made very cheaply from six or eight feet of nylon rope. The first eight feet of rope from Dad's boat anchor is about right.

There is an old bulletin in county extension offices of many states that shows how to make a rope halter for a cow or steer. Halters for sheep or lambs can be made the same way — by reducing the dimensions. The rope halter information was originally printed in a Cornell University 4-H bulletin and is probably available in many forms.

A halter is almost essential for working with junior show lambs at home, as well as around the fairgrounds. You want to train the lamb so that the halter won't be necessary for showing, but it will be helpful in getting the animal under control.

The best time to begin training the lamb is a matter of opinion and depends upon the size and experience of the junior exhibitor. A lamb that is not gentle enough for the kids to handle is not going to work, but one that is a total pet will probably go to sleep in the ring. For the smaller exhibitors I would rather have the pet than the runaway.

Leading with a halter is a good way to exercise the lamb and the easiest way to move it from one place to another.

Experienced exhibitors will say that you want the lamb nervous enough so that it will tense-up when the judge handles it. This makes the lamb feel harder. On the other hand, a lamb doesn't feel very firm when it's dragging its owner toward the barn.

Most lambs will tame down if worked with for about three weeks before the show. Activities such as washing, trimming, and carding will help a lot in this regard.

The most frequent question asked at our fair is, "Would it be OK for the smaller kids to show their lamb with a halter?" This is debatable, I think.

Everyone who has been around sheep shows knows we don't show lambs with halters. We show cattle with halters, but we hold lambs by the skin under the chin. Why is that? Who knows?

Of course, one should have the animal trained so it can be handled without a halter; but young exhibitors can become very upset if their lamb gets away in the ring.

One thing that helps with a lamb that is difficult to control is to walk it around to wear it down a little before going into the ring. This may result in a lamb that is too relaxed to show at its best; but if you're worried about holding on, maybe a relaxed lamb isn't so bad.

It is traditional to set the lamb by picking up the feet and placing them back down in the proper position. Again, we have a problem for the smaller kids with big lambs. While the kid is at the back end reaching for a rear hoof, the front end is heading for the barn.

If the exhibitor will put a knee in the lamb's chest and throw it off balance or pull the lamb forward a few inches, the lamb will move a rear foot that is badly out of position and very often stand the way we want it. A lamb that has been worked with will set up fairly well if the exhibitor pushes it one way or the other from the front end.

Once you have the lamb at the show, it's best to maintain the same ration you were feeding at home, except you will want to cut back on the amounts. The level of feeding on show day depends upon the individual animal and the amount of fill needed for it to look its best.

Some exhibitors will hold feed and water the morning of the show. A light feeding and control of water consumption the day of the show is generally the best practice.

Checking-out the show ring ahead of time is another important task. If there are high spots or holes in the ring,

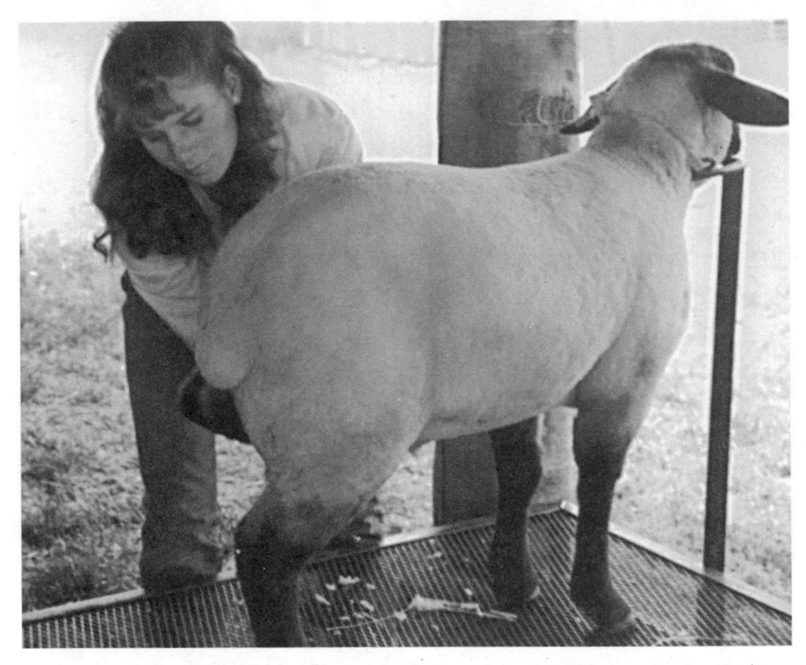

Working with the lamb on the trimming stand helps gentle the lamb and makes it easier to handle during the show.

remember that animals always look best when the front is higher than the rear. You don't want your lamb's feet in a hole just when the judge is deciding if it should be Grand Champion.

Know when your class will show and be ready when the class is called. You don't have to be first into the ring, but a exhibitor is more confident if ready to go when the class is called.

Part of being ready is having your clothes on. Exhibitors should change into their showmanship clothes well ahead of the time they will enter the ring to avoid the last minute panic this often causes. Junior exhibitors can wear coveralls or Dad's old shirt over their clean clothes while performing last minute chores.

The first rule of lamb showmanship is to make the lamb look the best it possibly can. The second rule is to watch the judge and be ready for instructions.

To me this says the appearance of the lamb should be the first priority. While it's important to watch the judge, the exhibitor must pay constant attention to the way the lamb is standing.

Making the lamb look its best means having all four feet under the animal and properly set. It means holding the lamb's head up and stretching the neck enough to give the appearance of length and size, as well as tightening the back muscles when the lamb is handled.

I've heard all kinds of ideas about bracing the lamb when handled by the judge. Talk with local 4-H leaders and judges to find out what is expected at your show.

Market lamb classes generally begin with lambs being led into the ring and circling clockwise. Often lambs will be led in and formed into a line before the judge asks the exhibitors to circle the ring. The size of the class as well as a judge's preference may affect this.

It's important to begin showing when entering the ring and continue showing until the class is dismissed. Even if it

Having a well-trained lamb gives the exhibitor confidence and permits watching the judge.

appears the judge isn't watching when lambs are brought into the ring, he probably is. Take every opportunity to make a good impression.

Don't become discouraged if you appear to be near the middle or the bottom of the class. The show isn't over 'til it's over. I've seen instances where exhibitors didn't know which end of the class was the bottom, and started relaxing when they should have been showing harder.

Remember, the goal is to make the lamb look good. Practice at home will produce the confidence needed to handle the lamb properly and make it easier to remember all of the things that are expected in the ring.

Wool Care

Older readers may remember a cartoon done by the late Ace Reed back in the early 1970's. The cartoon shows old Slim opening the mailbox and saying, "Hot dog! We got the check fer our wool! Now we can go to the cattleman's convention."

How true it is: The wool check always comes in handy, but it seldom gets the respect it deserves.

The popularity of sheep for small farms and junior livestock projects dictates that a significant portion of the wool harvested in the U.S. comes from small flocks. While larger producers are well aware of the need for proper care of their wool clip, owners of a few sheep should also do their best to harvest a quality product.

What can be done by the junior livestock producer or small flock owner to improve the quality of the wool marketed? We could start by finding a good shearer. There aren't many good shearers in some areas, and it's often necessary to pool flocks to permit hiring a shearer from some distance. Many wool pools sponsor shearing days for this purpose.

Plastic twine should not be used to tie fleeces, wool bags, or anything else near the shearing floor. Some producers are making it a policy not to use plastic twine anywhere on the farm, because of wool contamination from this material.

Wool should never be stuffed into plastic feed sacks or garbage bags, for the same reason. Plastic contamination of any kind will result in rejection of the entire lot.

Fleeces should be rolled flesh-side-out and tied with paper twine made for this purpose. Fleeces should be packed in wool bags. This is another reason for pooling small flocks for shearing — to have enough fleeces to fill a wool bag.

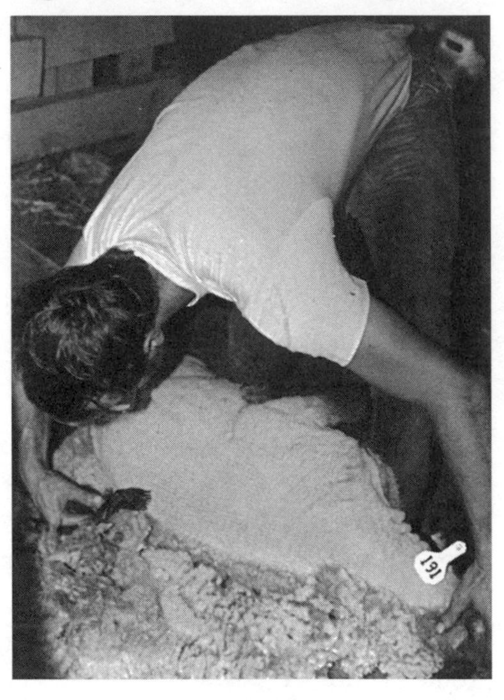

A good shearer is important for producing a quality wool clip.

Sheep must be dry when shorn, and shearing should be done on a clean, dry, surface. Fleece wool should be kept clean and bagged separately from tags or lamb's wool. Those who have enough sheep to use branding fluids for identification must be sure they buy scourable fluids only.

If both black-faced breeds and white-faced breeds are being shorn, the white-faces should be shorn first. This avoids contamination of these fleeces with black fibers, which may be left on the shearing head by the black-faced breeds.

The value of wool from white-faced breeds can be increased by skirting the fleeces and separating belly wool from fleece wool. Small flock owners with a few white-face fleeces may be able to market these to home spinners for a much better price than can be expected on the commercial market.

I should mention that not all white-faced sheep are created equal. Wool from some breeds is more desirable, and will bring a higher price for particular uses. Black wool (called natural colored by those who have it) is sometimes sold at a premium in the home spinning market.

Wool should be bagged and stored in a dry place until sold. It's best to put wool sacks on boards a few inches from the floor, to permit air circulation beneath them. Wool will draw moisture from concrete floors and rot.

Separating lamb's wool, tags, and fleece wool retains the value of the wool clip.

Then, of course, one has to sell the product. The common method of sale for small producers in many areas is through a pool arrangement where wool is sold cooperatively by a number of producers. If a pool is not available, the person with a very small quantity is at the mercy of whoever will buy it.

Whatever the marketing method, better quality will be worth more in the long term. The American Wool Council produces an excellent booklet on wool grades and types of wool produced by various sheep breeds. The booklet is called "Wool grades and the sheep that grow the wool", available from American Wool Council, Wool Education Center, 200 Clayton Street, Denver, Colorado 80206.

Most county extension offices also can furnish information on proper care and handling of wool.

SWINE

Selecting Market Pigs

Market pigs are an excellent 4-H or FFA project. The junior exhibitor who buys a good pig at the right size and a reasonable price is almost assured of a successful experience.

Even when things don't go well, market pig raisers can generally eat their mistakes. Few things in life are more reliable than that.

The first requirement for a market swine project is finding a pig of the right age and size for your show. If your pig is purchased from a breeder who is familiar with the show in question, he or she can often tell you what size and age of pig is needed.

Vo-ag teachers and 4-H leaders are good sources of advice, also. They know your feeding situation and understand that you might get different results than a swine breeder would expect on his own farm.

For many shows, market weight is around 230 to 265 pounds these days. This varies by area of the country, and should be checked locally before selecting a pig for a project.

Most market hogs will reach these weights around 5 to 5 1/2 months of age. There are variables, however. We must allow for slower growth during hot weather, a lack of pig feeding experience, and the week everyone has basketball camp.

I have devised the following table to help estimate growth rate for market pig projects. The table was taken from an older reference and tinkered with by my computer to adjust for faster growth expected from today's hogs.

GROWTH RATE OF SWINE

Age		Weight	
Days	Weeks	Pounds	Gain per Day (lb.)
7	1	4	NA
14	2	8	0.57
21	3	13	0.71
28	4	18	0.71
35	5	24	0.86
42	6	31	1.00
49	7	39	1.14
56	8	48	1.29
63	9	58	1.43
70	10	68	1.43
77	11	78	1.43
84	12	90	1.71
91	13	102	1.71
98	14	115	1.86
105	15	128	1.86
112	16	142	2.00
119	17	157	2.14
126	18	172	2.14
133	19	187	2.14
140	20	203	2.29
147	21	219	2.29
154	22	233	2.29
161	23	248	2.14
168	24	262	1.86

If my computer knows anything about pigs, we can forecast a pig weighing 233 pounds at 22 weeks of age. By counting backward we can calculate a pig about 50 days of age and weighing approximately 40 lb. should be about right for a show 110 days away. Some pigs will grow faster than these estimates and some will grow slower.

The table shows that a couple of weeks of age can be critical when choosing a pig for a particular show date. A pig that is too young may not have a chance to make market weight; and one that's too old may be difficult to hold within

At this stage the best predictor of these pigs' performance is the breeding behind them.

the desired weight limits.

Most junior livestock members prefer to select a pig that can be expected to reach market weight a week or 10 days before the show. Then the pig is held to a slower gain for the final portion of the feeding period. It's always easier to hold one back for a week or two than trying to stuff him the last week before the fair.

In addition to meeting the show's weight requirements (if there are any), the exhibitor would like a pig to finish at the proper weight for the animal's frame. This will narrow things further to only a 10 to 20 lb. range, in which the pig will be at its ideal show weight. Pigs should be weighed often to see if they are gaining as expected.

For those who don't have access to scales, weigh tapes are available to help estimate a pig's weight from heart-girth measurements. These are most helpful when they can be used in conjunction with an occasional visit to the scales.

How much should you pay for a junior pig project? Something above market price seems appropriate to me. The breeder deserves more than market price for giving junior exhibitors a chance to select one of his or her top animals.

On the other hand, nobody's animals are worth double the market price in my opinion. Youngsters who pay twice as much generally place no better than those who find a more economical source of pigs. I've seen that proven many times.

Everyone knows that 4-H and FFA exhibitors operate in an artificial market; but there has to be some economics in these projects, too. Paying a reasonable price for project animals is an important step toward keeping some perspective in the junior livestock business.

It takes a good pig at a desirable show weight to go home a winner.

Feeding Market Pigs

Feeding market pigs should be easy — like bathing a fish, oiling an eel, or bending a snake. But easy as it seems, producing a quality show pig at the desired market weight for a particular show date is far from automatic.

The first concern for a new pig owner should be worming of the pigs. This is normally done soon after pigs are obtained from the breeder; and a second worming 30 to 60 days later may be advisable in some instances. Your veterinarian is a good source of advice on the best time for worming and the most effective drug to use.

Ask your veterinarian about vaccinations, too. Disease problems vary by locality, and the local vet will know which diseases you should be concerned about. Purchasing pigs from a breeder with a well-planned health program can also help in this regard.

Now, let's tackle the question of feed for a pig. There are two common methods for obtaining a pig ration: 1. Buy a complete prepared ration, or, 2. Mix home grown (or purchased) grains with a prepared protein and mineral supplement.

The choice between these options depends upon the availability and price of good quality grain and prepared supplements versus the cost and quality of the complete prepared feed. You can obtain good results from either system, if you have quality ingredients in the home-mixed ration, or if you buy a good quality prepared feed when purchasing a complete ration.

You will notice some "ifs" in the previous sentence. The quality of the protein and the amino acid balance is very important in a pig ration. We may buy a feed with 14 percent crude protein, but whether the feed contains the amino acids needed by the pig depends upon the sources of protein in the feed.

The total protein level doesn't tell the whole story. If one or more amino acids are deficient, the cheapest feed may be the most expensive in terms of cost versus results. Personal experience and the experience of others must also be considered to determine the performance one can expect from a particular feed.

A combination show feeder and tack box that can be made in a vo-ag shop. This gets the tack out of the aisles and makes a neater exhibit.

On farms where home grown feeds and good mixing facilities are available, a ration that utilizes these feeds is usually the most economical. Where corn, barley, or wheat are available, these grains often form the base for home-mixed rations. When rations are mixed at home, ingredients must be thoroughly combined, and supplements containing antibiotics or other additives must be mixed and used properly.

Caution is in order when purchasing complete rations, too. Some feeds contain low levels of medication for younger pigs and may require a withdrawal period before slaughter. Read feed tags carefully.

It is generally recommended that pigs under 50 pounds receive a ration containing 18 to 20 percent crude protein; pigs 50 to 100 pounds receive a ration with 16 to 18 percent crude protein; pigs 100 to 160 pounds receive 16 percent protein; and pigs over 160 pounds receive a 14 percent protein ration.

These are averages that must be taken with a grain of salt, remembering what was just said about the hazards of looking at total protein percentages. A ration with more than the recommended protein level won't hurt the pigs, but generally costs more. This is why we reduce the protein level as pigs become larger and their protein need changes.

Should you hand feed the pigs or put them on a self-feeder? I think this is more of a concern to the people than to the pigs. Pigs will do fine on a self-feeder, if it is built to prevent wastage. Most junior exhibitors prefer to hand feed to control intake the last two weeks before show. This gives

A self-feeder is a good way to feed market pigs.

an opportunity to trim up the pig's belly by reducing his consumption. (Livestock shows cause us to do funny things.)

If the pig is becoming too large or overfat before show there are three ways to hold him back: Reduce the amount of ration fed, reduce the energy level of the ration, and increase exercise. A combination of all three is probably best. Then, we need to weigh the pig often to be sure we are getting the expected result.

Once a pig gets beyond 150 lb. we can normally expect 1 lb. of gain for each 3 to 4 lb. of ration consumed. A pig that is gaining 2 lb. per day will be consuming approximately 6 to 7 lb. of finishing ration per day.

"Consuming" is a key word here. It's possible to give a pig twice that much feed and have him eat half and roll around in the other half. If the pig seems to be eating 15 lb. of feed each day, you can bet he's being less than frugal with his ration.

Cutting the ration severely can reduce gains to near zero, but will make the pig very irritable and hard to confine. Feeding a bulky ration, such as oats, helps reduce gains and will

An automatic watering device must be securely fastened, or the pigs will take it apart and build themselves a swimming pool.

keep the pig in a better mood than a severe cutback in total ration.

I have been told that 4 lb. of oats per day will keep a 220 lb. pig at zero gain for some time; but like others who attend livestock shows, I've been told a lot of things. Using the scales is the best way to determine what works with your pig.

If a scales is not available, the following formula will help estimate the weight of a pig. Such measurements are not totally reliable, however, and should be used in combination with an occasional trip to the scales to check accuracy.

The pig feeding facility should include a system of providing fresh water at least twice a day and for keeping water available at all times. A self-serve watering system giving access to cool water is best, especially during the summer months. Shade is also important.

Heart Girth and Length Measurement for Hogs

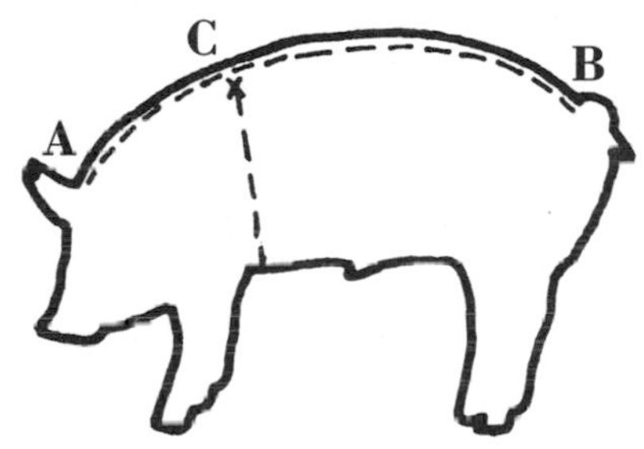

1. Measure the hog's heart girth in inches, at a point slightly behind the shoulders (**C**).

2. Measure length of body from the poll (between the ears), over the back to the base of the tail (**A-B**).

3. Multiply heart girth X heart girth X length and divide by 400 for estimate of weight in pounds.

Heart girth X heart girth X body length ÷ 400 = weight in pounds

Note: For hogs weighing less than 150 lb, add 7 lb to the weight obtained from the formula. For hogs weighing 151 to 400 lb, no adjustment is needed.

Source: The Stockman's Handbook, M.E. Ensminger

Fitting and Showing Pigs

Junior swine exhibitors generally agree that pigs are easy to prepare for a show, but difficult to handle in the ring. There's something about taunting a 250 pound animal with a whip or cane that doesn't engender confidence.

The best way to conquer this lack of faith is to begin training soon enough that the exhibitor has some assurance he or she is more proficient than the pig. While experienced exhibitors may begin training pigs 2 to 4 weeks before the show, a longer period of practice might be helpful for younger exhibitors.

Training should begin with getting pigs used to the handler and the cane or whip to be used in the show ring. Early training is also a good chance to learn how the pig responds to movements of the exhibitor.

Most pigs learn fast and respond readily to the show cane or whip. The pig should be tame enough to handle well, but not such a pet that it won't drive properly or respond to a tap on the hocks.

In some parts of the country everyone uses a show whip for driving pigs. In other areas the cane is still popular. Some hold their cane by the crook end and turn it over only when breaking up a fight. Where I grew up we were taught to hold the cane upside down and use the crook end for directing the pig.

Regardless of what the exhibitor uses for driving pigs, the show superintendent should be sure there are some adults in the ring to keep the animals from fighting. These people need a small piece of plywood (called a pig panel) to separate pigs.

The whip or cane should be used to start and drive the pig, but must be used judiciously. Check your 4-H and FFA manuals for directions on how to drive the pig. It helps to have a few practice sessions at home with someone playing the part of the judge. The ability to move and stop the animal can only be achieved through repetition.

This exhibitor has the pig between him and the judge and is doing a good job of watching the judge.

The way pigs are handled during training, loading, and weighing can have a major effect upon how they react during the show. We want to avoid undue excitement or anything that teaches the animals bad habits. It's best not to fight with them.

Many years ago I saw a 4-H leader put a five-gallon bucket on a pig's head and back the uncooperative porker into the show ring. This is much better than the squealing, beating, and mugging often seen in the hog barn.

The bucket trick works for loading pigs as well as moving them from place to place. The handler can aim the pig's tail in the preferred direction and follow him anywhere.

Fitting for show generally begins with washing. Most pigs need only a couple of washings a week or so before the show, if they are kept in a clean dry pen after washing.

Others might need washing a month or so before the show and again during the last week or two. This generally depends upon how clean the pig's pen is kept and whether a mud hole has developed.

Pigs are normally bathed again several hours before show time, or washed completely the day before the show and spot-washed on show day. A well-bedded pen can save a lot of extra scrubbing.

A good brushing after washing and again before showing will help the pig's appearance. A light spray of water from a spray bottle and a good brushing just before show will add shine to the hair coat. Be sure the pig is dry before going into the ring, however.

Keep water out of the pig's ears when washing, and rinse the pig thoroughly after washing to be sure soap is not left on the skin. Frequent brushing is a good way to improve the hair coat, as well as helping clean the pig.

It's easy to forget that pigs should be clean for a market swine class, as well as for showmanship. A clean pig always looks better than a dirty one, and nobody likes a show where the pigs look like they just walked in from the barn lot.

Market hogs generally won't need hooves trimmed unless there is excessive growth. If hooves need trimming, this should be done at least two weeks before the show to allow time for healing of any tenderness or mistakes in trimming.

Many exhibitors clip hair on the pig's ears two or three days before the show. Judges often suggest this be done a week before show to avoid the "fresh haircut" look.

Some judges suggest clipping hair both outside and inside the ear, while others prefer leaving the hair inside the ear. The author prefers the latter, as hairs within the ear help exclude dirt and protect the ear.

Hair should be trimmed with scissors or small hair clip-

Electric hair clippers are fine for clipping ears and tails. If the clippers have a cord, make sure it is out of reach of other pigs as well as the one you are trimming.

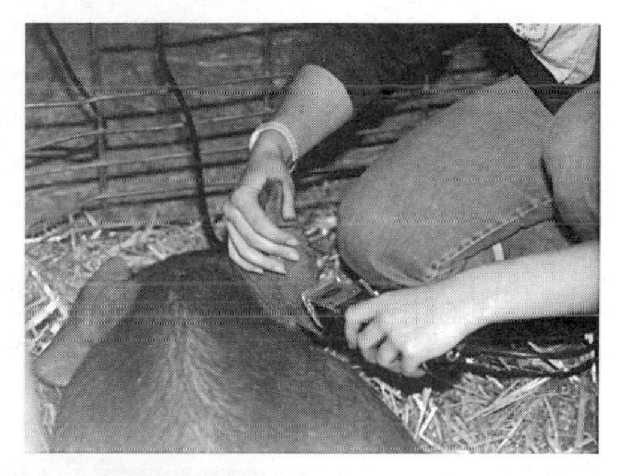

The pig's ears are clipped outside and along the edge of the ear. Hair should be left inside the ear to protect the ear.

The tail is generally trimmed from the base of the tail to the brush. This may vary in different parts of the country.

pers. If electric clippers are used, many pigs will bite the electric cord at the first opportunity, creating a dangerous situation. The cord can be protected by inserting it through a split rubber hose. Battery operated clippers are ideal, and slightly less expensive than the pig.

Hair may be clipped from the upper third of the tail or

all the way down to the switch, depending upon what part of the country you are in. The underline is trimmed from just above the teat line — down and under the belly.

In many areas the entire underline isn't trimmed, but only long hairs on the belly are clipped. Excessively long hairs around the nose, eyes, and mouth should be clipped, also.

Most shows discourage the old practice of using mineral oil and baby powder on show pigs. Meat packers have expressed concern about getting the hair off these slick critters. The powder and oil rule varies with locality and should be checked with your vo-ag teacher, county agent, or show officials.

Many exhibitors prefer to feed the pig half a normal feeding the day of the show. This feeding should be given two hours or more before show time. Water should also be restricted.

We don't want the pig too full, but you don't want him so hungry (or thirsty) that you can't get his attention in the ring. A pig that's too hungry is more likely to start a fight.

Withdrawal times must be observed for any medications used in feed or given as injections. Sometimes the feeds formulated for young pigs can not be fed close to slaughter.

A well-fitted pig handled by a good showman.

Breeding Swine Projects

Raising feeder pigs can be a rewarding project for junior livestock members. A breeding swine enterprise requires dedication and attention to detail, however. Keeping the investment low is always important, and today's swine industry demands close attention to disease prevention practices.

While facilities for breeding sows need not be elaborate or expensive, the potential for transferring diseases from one farm to another makes owning your own a boar advisable for most situations. This means the junior livestock member may need to own several sows to help defray the costs of buying and keeping a boar. Then, if you raise replacement gilts on the farm, it will be necessary to buy a new boar each time you keep some gilts, to avoid breeding gilts back to their sire.

These thoughts shouldn't discourage the beginning hog producer; but it helps to consider them when estimating costs and planning facilities.

It generally isn't necessary for junior livestock members to spend a lot of money for purebred breeding stock. Good crossbred gilts can often be purchased from commercial hog producers at a relatively low cost.

Crossbred sows are generally more prolific and better mothers than purebreds and are probably a better choice for most beginning hog raisers. Purebred boars, on the other hand, are often recommended to assure more uniformity of the pigs produced.

State and local swine breeder's associations schedule sales throughout the year where breeding stock is sold. Check with your local county agent for sale dates or other local sources.

Production records are valuable for making selection decisions. Both boars and gilts should be selected from sows with good production records over a period of years and from litters of ten or more pigs.

Gilts should be structurally sound with good feet and legs and should have a minimum of 12 functional, well-spaced nipples. A moderate amount of muscle is desirable, but excessive muscling should be avoided. Very heavy-muscled sows are seldom good mothers.

Boars and gilts reach sexual maturity at approximately eight months of age. Breeding them earlier than this age usually results in small litters. Gilts should farrow their first litter at one year of age.

When producing pigs for junior livestock shows, we generally would like sows to farrow 20 to 24 weeks before the show. This varies with each litter's genetic potential as well as the feeding program.

The gestation period for hogs is three months, three weeks, and three days. This means that sows bred 9 1/2 months before the expected show date will farrow at approximately the right time. Timing of these farrowings is important, as just a few weeks one way or the other can put pigs out of the market for a particular show.

Sows will generally be in standing heat three to four days after their pigs are weaned, regardless of the age of the pigs. Pigs are usually weaned at six to eight weeks. The sow can be bred at this time or held one or two more heat cycles if being bred to farrow for a particular show date. Heat cycles average 21 days.

Sows must farrow twice a year to be profitable. If there are several local or district livestock shows in your area, it may be possible to breed sows to raise pigs for more than one show each year. A certain percentage of your pigs will have to be sold as feeders or fed for market rather than going to a livestock show.

The sow's greatest needs are shade and fresh water in the summer and draft-free shelter in the winter. Single unit 'A' frame houses faced away from prevailing winds provide adequate shelter for adult hogs during the winter. Similar units with doors on both ends and built-in stalls work well for farrowing units.

Following a few simple rules will go a long way toward maintaining a disease-free herd. These include:

1. Buy healthy, disease-free breeding stock.
2. Isolate new hogs for at least 21 days after purchase.
3. Don't let "outside" hogs onto the farm.
4. Be careful of visitors, and don't visit other hog operations in shoes and clothing worn around hog facilities at home.
5. Move hog lots occasionally.
6. Don't let mud holes develop.
7. If using individual farrowing houses, clean and move them between farrowings. Don't farrow on the same site more than once in two years.
8. Keep facilities and equipment clean.
9. Maintain a good nutrition program.
10. Watch animals daily and treat immediately if problems appear.

Feeding Breeding Sows

Feed costs are a major concern for any swine enterprise. A well-planned and economical feeding program is just as important for 4-H and FFA members with a few sows as it is for large commercial producers.

The first principle in feeding sows is to consider the stage of production. When a sow is lactating, her nutrient requirements may be several times her needs during early gestation. Rations must be adjusted throughout the year to maintain health and productivity in the sow, as well as solvency in the bank account.

Balancing a ration for hogs is more difficult than for ruminant animals, such as cattle and sheep. Because swine have no digestive vat, such as the rumen in cattle, there is no bacterial breakdown of feeds before digestion in the true stomach. This means hogs are unable to digest large quantities of roughage, as ruminants do.

Hogs also need complete proteins with the proper balance of amino acids, whereas the ruminant animals can reassemble amino acids and manufacture proteins through bacterial action in the rumen. Hogs require B vitamins in the diet, while ruminants get these from bacterial activity.

All of this means rations for breeding swine must be well balanced and provide amino acids and vitamins in the proper amounts. It also says that many feeds formulated for ruminants are not good for hogs; and certain additives to cattle and sheep rations, such as urea, can be toxic to hogs.

One solution to the balanced ration question is to buy a complete feed that includes grains for energy and has all of the necessary amino acids, B vitamins, and minerals. A second alternative where good quality grains are available is to feed a commercial protein supplement and mix it with grains or other feeds that can be found at a good price. These supplements give directions on the tag for mixing and feeding to various ages and classes of animals.

Grains commonly fed to hogs include corn, barley, wheat, and oats. Triticale is an excellent swine feed where it is available. Moldy grains must be avoided for breeding swine rations. Certain grain molds produce toxins that can affect egg production in the sow or sperm production in the boar.

Grains are commonly rolled or ground for swine rations, but some grains, such as wheat or triticale, may be fed without grinding under certain conditions. Grinding gives better feed efficiency with most grains, but the cost of grinding can exceed the benefit in situations where whole grains are cheap and grinding is inconvenient or expensive.

Feed requirements for the sow are greatly increased during lactation.

Breeding sows may need up to 15 or 20 pounds of feed per day during lactation to keep them milking and to put them in condition to breed back after the pigs are weaned. There is no room for high fiber feeds or fillers during this stage of production. Condition of the sow and the number of pigs suckling should help determine the amount fed.

Amino acid levels and B vitamins are important at this time as are the high energy grains. If grains are mixed at home, a commercial protein supplement should be fed to assure the proper balance of amino acids, B vitamins, and minerals.

The sow also needs a balanced diet during gestation, but her nutrient requirements are greatly reduced during this stage. Sows may need only four or five pounds of grain ration per day during early gestation, and care must be taken to prevent sows from getting over-fat during this period. Some researchers suggest the sow's ration should be increased to about 10 pounds per day two weeks before farrowing. The amount of feed needed at all stages varies with size and age of the sow.

Because small amounts of feed won't keep the sow satisfied, high-fiber feeds are often used as fillers for early gestation rations. This is sometimes provided by legume pastures or alfalfa hay. Higher fiber grains such as oats are also good gestation feeds. It is often best to hand-feed gestating sows because of the smaller amount of feed needed.

Cull fruits and vegetables can be fed to swine, but such things as raw beans contain toxins and should not be fed to hogs. Raw potatoes are not digestible to hogs, but cooked potatoes can be fed. Root crops such as beets or carrots are digestible and provide some vitamins but little energy because of their high water content.

Garbage is not a feasible feed source for small numbers of hogs because of licensing and cooking requirements. Cattle feed should not be fed to hogs because of the possibility of ingredients such as urea or other additives that may be toxic to hogs. If the total ingredients of cattle or sheep feeds are not known, they shouldn't be fed to hogs.

Salt, trace minerals, and vitamins should also be included

in swine rations. Let's not forget to provide clean water. Automatic waterers are best and should be leak proof or on a base of some sort to prevent the sows from making a large swimming pool.

Baby Pig Care

Most junior swine showmen buy feeder pigs from others for entering in junior market pig classes a few months later. There is room, however, for those who prefer to raise pigs from their own sows and sell the excess to other 4-H and FFA members.

Don't forget though: There won't be any excess, if newborn pigs aren't taken care of. Young swine producers learn very quickly that preventing death loss is the key to success in this type of operation.

Sows producing pigs for junior livestock shows are generally bred to farrow five to six months before the show. For many this means farrowing in the winter and early spring months. Pigs are very sensitive to cold for the first few days after birth, and special care is required to prevent death loss this time of year. The first few days after birth are the most critical period for preventing these losses.

Ideal temperatures for pigs are 90 to 95 degrees F. for the first four days after birth, 85 to 90 degrees from five days through 10 days, and 80 to 85 degrees from 11 to 15 days of age. After two weeks of age the pig is able to regulate his body temperature and can adapt to much cooler temperatures. Prevention of drafts is also important to pig health.

Most junior livestock raisers don't have the luxury of an environmentally controlled farrowing barn, and may have to improvise to provide supplemental heat to the pigs. Small producers often have to compromise on the ideal temperatures for the sow when farrowing in a corner of the barn or in an outdoor hog-house.

Even in enclosed facilities it is better (and cheaper) to warm the pigs without providing too much heat to the sows. Many pig raisers accomplish this by providing supplemental heat to the pigs through the use of heat lamps, heat pads, or hovers.

A University of Illinois study showed a reduction in baby pig losses from 13.4 percent to 4.4 percent when heat lamps with 250-watt bulbs were suspended 18 inches above the piglet sleeping area when the farrowing room was kept at 70 degrees F. (Heat lamps must be properly protected and used carefully in a hog barn.)

Clean, dry quarters are important to pig health.

One of the most economical ways to help pigs maintain their body temperature is to provide plenty of straw bedding. A second Illinois study proved that when the farrowing room temperature is 50 degrees F, covering the floor with four inches of straw is equivalent to raising the environmental temperature to 65 degrees.

Bedding reduces the pigs' heat loss to the floor, and burrowing into the straw also lessens heat loss. This type of bedding should be provided in an area where the pigs are protected from being laid-on or stepped-on by the sow.

Mentioning heat lamps and straw in succeeding paragraphs makes me a bit nervous. The importance of securing lamps and protecting them from the sow cannot be overemphasized. Placing the lamp at the proper height is also critical to fire safety.

The ability of the pigs to burrow into straw or similar bedding means the ideal temperature at floor level can be reduced and lamps and hovers can be placed at a greater distance from the pigs and bedding. A thermometer should be placed in the sleeping quarters to determine the temperature at the pigs' level.

If the pen area for the sow is bedded with straw, this material should be chopped if it is long enough to interfere with movement of the pigs. When farrowing crates are not used, baby pigs should be protected with guard rails 8 inches above the bedding and 8 inches from the wall of individual farrowing pens.

Observing the pigs is important. Pigs that are chilled may shiver or huddle together under the heat source. When the pigs are too warm they may sleep away from the heat lamp or hover.

Survival of very small (runt) pigs is never as good as pigs of normal size. Studies have shown that nearly 60% of newborn pigs weighing less than 2 pounds will die under normal conditions.

Researchers have been able to cut losses of these small pigs with supplemental feeding of a commercial milk replacer or a mixture of 1 quart whole cow's milk, 1/2 pint "Half & Half", and 1 raw egg. This is administered in 15 to 20 ml. doses once or twice per day with a soft plastic tube attached to a syringe. This is time consuming, but probably worth it for small numbers of runts.

Sow's milk provides all of the nutritional requirements

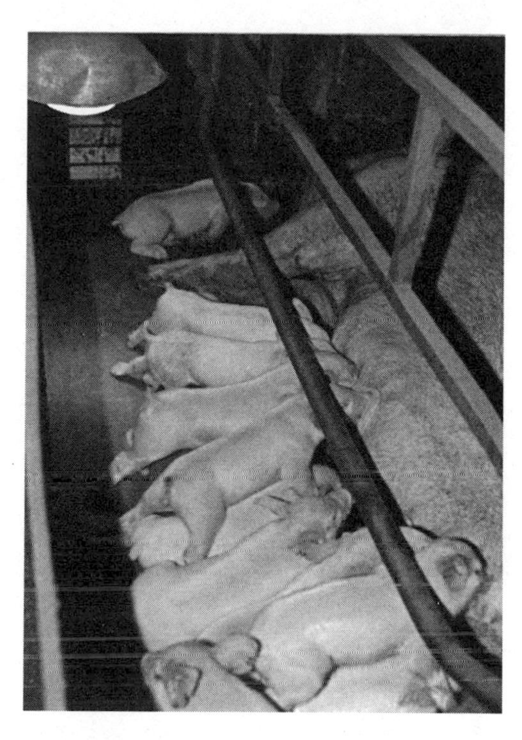

Farrowing stalls are great if you have them.

for newborn pigs, except iron. Iron deficiency is generally prevented through injections at 3 or 4 days and again at 2 weeks of age. Iron is available in the soil, and we used to suggest providing indoor litters with some clean dirt (a seeming contradiction in terms). Iron injection is more common nowadays.

Pigs should be offered dry feed at 1 to 2 weeks of age, and a good quality starter ration should be provided in a creep area. By the time pigs are 3 to 4 weeks of age, milk production from the sow begins to decrease, and pigs should be eating adequate amounts of starter ration in order to grow rapidly.

Pigs that are eating well can be weaned as early as 3 to 4 weeks. The younger the pig, the more stress and risk associated with weaning. The Pork Producers Handbook available through agricultural universities suggests weaning over a 2 or 3 day period, weaning the larger pigs in the litter first. This handbook suggests only those pigs weighing 12 pounds or more should be weaned.

DAIRY CATTLE

Selecting a Dairy Heifer Project

I'll never forget my FFA dairy heifer project. Her name was "Such and Such, Something or Other, Snowball." I remember the Snowball part because she was nearly all white, like a snowball. She was built like one, too —just as round and smooth as she could be.

I thought she looked pretty good (nice and round and fat), and the judge at the county fair liked her OK, too. He placed her second in a class of four. Looking back now I can see that the judge and I were both wrong, unless that happened to be a pretty lousy class of heifers.

That heifer just continued to get bigger and rounder until calving time, and then we could see there was a problem here. She was too round — and had very little dairy character or femininity. Her udder finally developed into the size and shape of a volleyball; and she never gave over 40 pounds of milk the best day of her life, which wasn't all that long.

This wasn't exactly what I had in mind; but I kept her two years, bred her to the biggest bull in the stud book, and had some nice visits with the vet each calving as we tried to pull the biggest bull calves you ever saw out of that ungrateful heifer. One of the calves lived, too, just to prevent a total loss from the enterprise.

So why am I giving advice on selecting dairy heifers? Well, it's like your dad says, "So you won't make the same mistakes I made." If you're going to make mistakes, at least they should be new ones.

Selecting a dairy calf, especially the first one, can be a real challenge to a person's judgment and foresight, because most of them are chosen as young and immature animals. Attention to good type is both practical and essential in the selection of animals for showing, or in the establishment of a foundation herd.

The structurally correct animal will always catch the judge's eye, and animals of desirable type generally produce and reproduce for a longer period of time.

A beginner may ask what we mean by "type." Type is

A show heifer of desirable type.

the general form, structure or character of a particular animal. Simply put, it is the way an animal is put together — its conformation and eye appeal.

When selecting animals for show, the exhibitor needs to learn the qualifications for the breed and to select animals that, when grown out and fitted, will be attractive to the judge.

Except for differences in color, size, and head character, all dairy breeds are judged on the same standards, as outlined in the Dairy Cow Unified Score Card. Anyone showing dairy cattle should be familiar with the unified score card and the criteria it suggests for conformation.

Where do we begin with selection? First, you'll have to choose a breed, and that might be easy if you already know which breeds you like. Or you may have a particular breed that is readily available. The breed chosen is important, so consider all of the factors before you buy.

One shouldn't buy a certain breed just to be different or to avoid stiff competition in the show ring. Exhibitors will learn more if there is strong competition in both numbers and quality of animals. A good dairy person will generally be successful with any breed.

The next question is whether to buy a purebred or grade animal. This decision will be greatly influenced by the ultimate objectives. Are you starting a foundation herd, or are you showing at breed association shows as well as youth shows?

Remember, that registration papers don't assure that your calf will be the kind of cow you want. My FFA heifer was registered, but that didn't help her much.

Good calves can be found in most parts of the country, but a little travel may be required to locate the calf you want in some areas. If possible, buy from someone you know has a high producing herd that is free of diseases.

If the decision is to go the purebred route, and this will be your first experience at raising a calf, there may be limits on where you can buy an animal. Some breeders won't sell to just anybody because of the chance of damage to the herd's reputation if the animal is not raised properly.

If the calf is primarily for show purposes, it is important to buy one that was born at the right time. Dairy cattle are shown in age groups, and primary consideration should be given to calves born near the beginning of their age class. A calf near the top end of its age class often has a big advantage in size and appearance over younger calves in that class.

It's best to select from calves that are at least three to

Holstein heifers at breeding age. Size for age is an important factor in selection of breeding heifers.

four months of age. There is less danger of death loss, and a more reliable evaluation of the calf's future can be made at this time than is possible at an earlier age. One way to determine if the calf is of desirable size for its age is to compare its heart girth measurement with the established normal measurement for that breed at a particular age. The chart, Heart Girth Measurements and Normal Growth of Dairy Calves and Heifers, accompanying this chapter gives expected heart girth measurements for calves of various ages.

Look for an alert heifer with good stature (size and scale) and length of body. She should be well-proportioned and have good dairy character, evidenced by a clean-cut neck and sharpness over the withers. The rump should be almost level and have good width. Body capacity will be shown by a wide chest and well-sprung, open, and deep ribs. The rear legs should be straight, as viewed from both the side and the rear.

If you don't have previous experience in selecting heifers, it would be well to take an experienced person with you to help make the selection.

Udders are very important to dairy cows. The old saying, "No udder, no cow," is very true in the dairy business.

Because it isn't possible to tell much about udders on young calves, you may want to ask to see the calf's mother and milking daughters of the calf's sire.

Size, shape, and quality of the udder as well as attachments are all important. Probably the best indication of a strong attachment is a good crease down the center of the udder, with teats hanging straight down or pointing slightly inward. The quarters should be evenly balanced and symmetrical. A good quality udder is characterized by a soft texture, is pliable and elastic, and is well collapsed after milking.

One other detail that may be important to you is color pattern. If an animal meets the color requirements of the dairy breed association, no discrimination against color or color pattern should be made at dairy shows.

If you have a calf with white knees, you'll find that manure stains on the knees are difficult to get rid of. Not a big thing, really. I would hate to encourage selection based upon the color of knees.

Don't forget the production records! After the heifer's show days are over (as a heifer, anyway) you want her to be a good producer and capable of having offspring with good production potential. Again, someone with dairy experience can be helpful in interpreting production records.

Selection will always contain a certain amount of guesswork, especially when buying young calves. However, a systematic approach combined with some knowledge and good judgment will reduce the guesswork and improve your chances for a good experience.

HEART GIRTH MEASUREMENTS FOR NORMAL GROWTH OF DAIRY CALVES AND HEIFERS

Age Months	**Holstein Brown Swiss** Inches	**Guernsey** Inches	**Jersey** Inches	**Ayrshire** Inches
Birth	30.0-32.0	28.0-30.0	24.5-26.5	28.5-30.5
1	33.5-35.5	30.5-32.5	28.5-30.5	31.0-33.0
2	37.0-39.0	33.5-35.5	31.5-33.5	34.5-36.5
3	40.0-42.0	37.0-39.0	35.5-37.5	37.5-39.5
4	43.5-45.5	40.0-42.0	37.0-39.0	41.5-43.5
5	47.0-49.0	43.0-45.0	40.5-42.5	44.5-46.5
6	50.0-52.0	46.0-48.0	43.5-45.5	47.0-49.0
7	52.5-54.5	48.5-50.5	46.0-48.0	50.0-52.0
8	53.5-55.5	50.5-52.5	48.5-50.5	52.0-54.0
9	56.0-58.0	52.5-54.5	50.5-52.5	54.0-56.0
10	57.5-59.5	54.0-56.0	52.0-54.0	56.0-58.0
11	59.5-61.5	55.5-57.5	54.0-56.0	57.0-59.0
12	61.5-63.5	57.0-59.0	55.5-57.5	58.0-60.0
13	62.0-64.0	58.0-60.0	56.5-58.5	59.5-61.5
14	63.0-65.0	59.5-61.5	57.5-59.5	61.0-63.0
15	64.0-66.0	60.5-62.5	58.0-60.0	62.0-64.0
16	65.0-67.0	61.5-63.5	58.5-60.5	63.0-65.0
17	66.0-68.0	62.5-64.5	59.5-61.5	64.0-66.0
18	67.5-69.5	64.0-66.0	60.5-62.5	65.0-67.0
19	68.0-70.0	64.5-66.5	61.5-63.5	65.5-67.5
20	69.5-71.5	65.0-67.0	62.0-64.0	66.5-68.5
21	70.5-72.5	66.5-68.5	63.0-65.0	67.5-69.5
22	71.5-73.5	67.0-69.0	64.0-66.0	68.0-70.0
23	72.0-74.0	68.0-70.0	65.0-67.0	69.0-71.0
24	72.5-74.5	69.0-71.0	65.5-67.5	70.0-72.0
25	73.5-75.5	69.5-71.5	66.0-68.0	70.5-72.5

Source: Extension Mimeo 2556 prepared by B.F. Kelso, Extension Dairy Scientist, Western Washington Research and Extension Center, Puyallup, WA.

Care and Management of the Dairy Heifer

The majority of 4-H and FFA dairy project members are from dairy farms. They may have a ready supply of good quality dairy animals, and their parents know how to feed and care for heifers.

This isn't always the case, however. Dairy projects may begin with a purchased heifer from another farm, and some result from the offspring of the family milk cow. Regardless of the calf's origin, good management is required for the dairy heifer to attain her full potential.

Proper care of a dairy heifer isn't complicated, but feeding, housing, and management must all fit together to raise a healthy, productive heifer.

One of the first requirements after the calf is born, or after getting the calf from the breeder, is to positively identify the heifer and set up a system of records for important events, such as vaccinations, growth rate, health status, etc. Identification can be done by a sketch of the animal and its markings, or by photographing the calf from both sides and the front.

A more permanent identification is the tattoo. If the calf is to be registered with a breed association, be sure to check the association rules before tattooing.

The heifer must be vaccinated for brucellosis at the proper age, and the veterinarian will put a brucellosis shield in her right ear at that time. Dehorning should be done by one or two months of age, especially if the heifer will be shown at a fair. Most fairs won't permit showing dairy animals with horns.

Removal of extra teats should be done as soon as they can be distinguished from the four main teats. This must be done carefully. There have been cases of the wrong teats being removed. County extension offices in most states have information on removing horns and extra teats from dairy calves.

A calf doesn't need a fancy place to live, but she should have plenty of fresh air and a dry place to lie down. My old friend Eddie Thomason of Yakima, Washington used to tell 4-H members to get in the calf pen and sit down on the bedding for about five minutes. That's a sure way to tell how dry it is.

Calf hutches are a popular management system in many parts of the country.

Calves should be kept in individual pens until they are weaned. This will decrease the chance of spreading disease from one heifer to another and makes it possible to watch the calf's eating habits more closely. Penning individually also prevents the problem of calves nursing each other.

About a week after weaning, calves may be placed together in small groups. Of course, good sanitation practices must be observed whether calves are raised individually or in groups.

The management timetable listed below is a good reminder of approved management practices for dairy calves.

MANAGEMENT TIMETABLE FOR DAIRY CALVES AND HEIFERS

1. Disinfecting Navel: As soon as possible after birth, dip the navel cord in iodine or similar antiseptic.

2. Colostrum: Make certain the calf receives feedings of colostrum from its dam or pooled colostrum from more than one cow within 24 hours following birth.

3. Removal from Dam: At birth to three days of age, preferably as close to birth as possible, for better management of the dam's udder and the calf.

4. Identification: Identify by means of an ear tag or other means, such as tattoo, color marking sketch, or photo, as soon as possible after birth.

5. Dehorning: This is best accomplished when the calf is 10-14 days old.

6. Removing Extra Teats: Rudimentary or extra teats should be removed as soon as they can be distinguished from the four main teats.

7. Dry Feed and Water: Provide calves with calf starter or equivalent grain mixture with a few days following birth. Start forage feeding and provide water within a week following birth. Provide free-choice mineral feeding as soon as feasible.

8. Weaning from Milk Replacer: Wean at six weeks if calves are thrifty and eating at least 1.0 pound of suitable grain mixture per head daily. Continue milk or milk replacer feeding as long as necessary for unthrifty calves.

9. Removal from Individual Stalls or Pens: Wait until the calf has been weaned from milk replacer for one or two weeks.

10. Vaccination: Brucellosis or Bang's Vaccination should be done before six months of age. Four months is preferable to avoid positive blood tests showing up at a later age.

11. Checking Growth: Important periods for checking growth are at 30 days, 6 months, 12 months, 15 months, first freshening, and second freshening.

12. Breeding Heifers: Plan to have heifers at proper size for breeding at 13-15 months of age. Avoid over-conditioned or fat heifers as this may be detrimental to future production.

RECOMMENDED BREEDING SIZE
FOR DAIRY HEIFERS

Breed	**Body Weight** Pounds	**Heart Girth** Inches
Ayrshire	650-700	61-63
Brown Swiss	750-800	64-66
Guernsey	600-65	59-61
Holstein	750-800	64-66
Jersey	550-600	58-60

Source: Extension Mimeo 2556 prepared by B.F. Kelso, Extension Dairy Scientist, Western Washington Research and Extension Center, Puyallup, WA.

Feeding Dairy Heifers

A well-grown heifer that will become a productive cow is the primary goal of a feeding program for dairy heifers. Heifers of the larger breeds are expected to breed at 15 months of age and weigh 1,200 lb. at first calving. These heifers need to average 1.5 lb. of gain per day to reach the desired weight of 750 to 800 lb. at breeding. This leaves little margin for error in the feeding program.

When a 4-H or FFA dairy heifer project is purchased from another farm, the buyer should make sure the heifer has the size and growth that can be expected for her age. Project manuals for 4-H and FFA members often contain a chart showing the approximate weight and growth rate to be expected of heifers within the common dairy breeds. The heifer's weight can be estimated with a taped heart-girth measurement if scales aren't available.

Washington State University prints a two-page piece called "Management Timetable And Growth Standards For Dairy Calves And Heifers." This publication contains a chart for evaluating growth rate, as well as helpful management suggestions. I would expect county extension offices in other states have similar publications.

If the project heifer is from your own farm, the feeding program begins with proper care and colostrum feeding the first few days after birth. Calves generally receive whole milk or a quality commercial milk replacer from three days of age to weaning; but colostrum can be fed beyond the first few days if diluted with water at a ratio of approximately two parts colostrum to one part water.

The sooner a young calf begins eating dry feed the better. The calf should be provided with a small amount of high quality calf-starter (grain mixture) in a feed box within the first few days of age. A very young calf won't eat much dry feed for a few days, but the process can be speeded up by putting a small handful into the calf's mouth right after feeding the milk. Calf starters should contain at least 16% protein.

The calf should be provided with a calf-starter (grain mixture), good quality hay, water, and minerals at an early age.

Good quality alfalfa hay should also be made available within the first week, as should fresh water and trace mineralized salt. The calf will soon be eating small amounts of calf-starter, and can be weaned from milk replacer at 6 weeks of age if it is doing well and eating at least 1 to 2 lb. of starter per day. Heifers of the larger breeds should receive 4 to 5 pounds of calf starter and free-choice hay from 3 to 9 months of age.

For those who would prefer to mix a grain ration at home, rather than buying a commercial calf-starter, the following is a recommended grain mixture for young calves.

Coarse ground or rolled barley or corn	50 lb.
Coarse ground or rolled oats	40 lb.
Linseed or soybean meal	10 lb.
Total	100 lb.

One pound of trace mineralized salt and 0.1 pound of steamed bonemeal or dicalcium phosphate should be added to each 100 pounds of this concentrate.

Concentrates for calves under 9 months of age should contain a coccidiostat to prevent losses from coccidiosis; but should not contain non-protein nitrogen sources, such as urea. Urea can be used for part of the protein requirement in rations for older heifers.

Don't forget clean, fresh water! A heifer calf will drink 3 lb. of water for each lb. of dry feed consumed.

By the time the heifer is 6-months old, she can handle a lot of hay and some pasture. Pasture grasses often contain considerable water and may cause a pot-belly in addition to being inadequate in nutrition for heifers under 1 year of age. These heifers should receive hay and grain rations in addition to pasture.

Heifers more than 1-year of age will often receive 4 to 6 lb. of concentrate ration along with good quality hay or pasture. Because hay or other forage often makes up most of the ration for dairy heifers, the quality of this forage is important. This includes digestibility and energy factors, as well as protein content.

A feeding program for dairy heifers should keep them in good condition and growing properly, but we don't want to get them fat. Research has shown that over-conditioning of young heifers lowers their lifetime milk production potential. Heifers to be shown at junior livestock shows are often fed to a higher level of condition than other dairy replacements. Junior showmen must be careful not to over-feed these heifers to the detriment of their later production.

Show heifers are generally evaluated several months before show date to determine if they have the desired size and condition. If the heifer has been fed well and has the needed size and growth for her age, the major concern is making sure the animal is not too fat or too thin.

If the heifer is too thin, we increase the concentrate or grain portion of the ration. If she is too fat, we decrease this portion.

The amount of concentrate needed varies with the animal's size and individual needs, as well as the quality of the forage in the ration. Some yearling heifers may require little or no concentrate to achieve proper condition, while others may need up to five or six pounds of grain concentrate per day. The heifer calf will require a higher protein ration than a yearling heifer.

Fitting and Training Dairy Heifers

Proper feeding and training are two essentials for preparing a dairy heifer for show. The poorly conditioned animal won't look her best, and one that isn't well-trained is impossible to pose correctly.

Training the animal to lead should begin early. Most dairy calves are gentle and can easily be trained to lead when they are penned and handled on a regular basis.

Training begins with fitting the heifer with a leather or nylon strap halter and tying her to a wooden fence or wall where she can't get away. A solid wall is best to be sure the heifer can't get a foot caught or hurt herself. Feeding some hay on the ground where the heifer is tied will help keep her occupied.

Rope halters are not usually needed or recommended for dairy calves, because these halters can rub the hair. While the heifer is tied she can be brushed or groomed to get her used to being handled. The animal can be left tied during the day, or for just a few minutes at a time. Within two to three days most heifers will be gentle enough to begin teaching to lead.

Later the heifer can be led with the leather show halter. Some calves will resent the chain under the chin on the show halter, and this can be removed in the beginning if desired.

After the calf is leading well and can be taken out of the pen into a larger area, the exhibitor should begin teaching the heifer to walk slowly and in small steps. An adult should be

Training the animal to lead and stand properly should begin several months before the show.

there to help and can play the role of the judge while the exhibitor practices leading and posing the animal.

The heifer should be kept in a clean, well-bedded pen when she is not outside and should be penned the last two months before the show. Good bedding and a dry pen are the best prevention for manure stains.

Show cattle are usually kept inside during the day to prevent sunburning of the hair. Blanketing dairy cattle the last two or three weeks before show has been traditional in many areas. The modern trend is not to blanket — because this discourages hair growth. Exhibitors generally want longer hair on the animals to permit a better fitting job.

Washing can begin several weeks before the show and may need to be performed a number of times on hard-to-clean or stained areas. Because washing too often can remove natural oils from the hair, this job should not be overdone. The heifer should be brushed after washing.

Many kinds of liquid soaps are suitable for washing livestock, as are the special soaps sold for this purpose. The livestock soaps are made to lather in cold water and may work better for situations where warm water isn't available. All soap must be thoroughly rinsed out of the hair after washing.

It's important to keep water out of the animal's ears. The inside of the ears can be cleaned with a rag dipped in rubbing alcohol to remove wax and dirt.

Dairy heifers are clipped around the head, neck, tailhead, and tail as well as at particular points on the topline to give the straightness and sharp lines desired. The entire head should be clipped, except for the long hairs on the nose. These should be left because the animal uses these hairs to feel things like feeders and other objects (and they make the muzzle look wider).

The ears are clipped inside and out. The neck is also clipped and blended-in with the shoulder area. We want the effect of smoothness and sharpness around the shoulders and over the withers. Proper clipping will give this effect and longer hair may be left on top to give a sharper appearance when combed up.

An expert fitting job on a nice show heifer.

High points on the topline or around the tailhead are clipped to make a straight topline. The tail is clipped from a point a few inches above the switch up to the tailhead.

Clipping requires practice and experience. Younger 4-H members can learn by clipping a practice animal that won't be going to the show, and by clipping the less difficult portions on their show heifer. Older members should develop the ability to do the clipping themselves. An experienced exhibitor should be consulted for advice on clipping individual animals.

Much of the clipping can be done two or three weeks before the show. Particular areas, such as the head, should be reclipped a few days before the show. Getting the animals used to the clippers well ahead of the show is a good practice and can prevent those last-minute disasters so unsettling near show day.

Showing Dairy Heifers

The previous chapter described how to train your dairy heifer and prepare her for show. This one will discuss show-day preparations and what's expected in the ring.

The first task is to find out when your class will show, so you can be ready well ahead of time. You have been grooming and training this heifer all summer — and don't want to be eating corndogs when your class goes into the ring.

Confidence in yourself and your animal is essential for success in showing livestock. The best way to build confidence is being ready the day of the show and knowing what you want to do in the ring.

The cattle come first on show day. Dairy animals must be fed properly the day they are shown, and grooming tasks must be well-planned. Murphy's law says whatever can go wrong will happen 10 minutes before your class goes into the ring.

Many exhibitors will feed hay in small amounts and on a continuous basis the day of the show. The goal is to keep the animal eating to produce the desired fill at show time.

Some also use a bulky feed such as soaked beet pulp or silage to give extra fill on show day. Whether the extra fill is needed depends upon the individual animal and the appearance desired.

The amount of feed and water given is varied according to the individual animal and the interval between feeding time

A drink of water before showing is usually needed to provide the desired fill and appearance.

and show time. While cows may eat whatever is put in front of them, heifers will need to be conditioned to any special feeds that will be given at the show. (It's a good policy not to do anything at the show you haven't tried at home first.)

Water is generally held back for several hours before show to assure the animal will drink before going into the ring. Then the amount of water given is adjusted to produce the proper fill.

Animals often require some spot-washing the morning of the show, and the tail switch is washed and combed. The heifer has likely been washed the day before and should not need a complete bath on show day. Don't worry about what the other kids are doing, you don't want the heifer wet when your class is called.

One bit of excitement everyone could do without is the last minute clothes changing and subsequent rush to the show ring. One way to be prepared and keep the white clothes clean is to dress for the show in plenty of time, and then wear coveralls over everything until soon before going into the ring. If coveralls aren't available, Dad's oversized work-shirt is a good substitute. Dairy exhibitors are expected to wear white in some parts of the country, but this varies with local customs and requirements.

Before entering the ring, make sure the show halter is adjusted to fit the animal properly and the lead strap is coming out of the left side of the halter (the animal's left). You should be near the entrance to the ring when your class is called; but you don't want to be there too long ahead of time. If your heifer stands too long she may become tired in the ring.

Many junior exhibitors are reluctant to be first into the ring, but they shouldn't pass up this opportunity if it presents itself. The first animal into the ring often has the immediate attention of the judge and has an excellent chance to make a good impression.

Know as much as you can about your animal. The judge may ask the birthdate, the sire, or in the case of older heif-

For a younger exhibitor, it's nice to have a heifer that is about your size.

ers, the judge may want to know if she is bred and the expected calving date.

Dairy animals are shown with the exhibitor walking slowly backward. Practice before the show will help determine the walking speed at which the heifer looks her best. The lead strap should be coiled in the exhibitor's hand but not wrapped around the hand.

The animal's head should be held high. The exhibitor must keep an eye on the judge, but not to the extent of forgetting how the animal is walking or standing. When the judge motions to stop, the exhibitor sets the animal.

Dairy heifers are posed with the rear leg nearest the judge a step backward from the other rear leg. Cows are posed with the rear leg nearest the judge forward. Most animals can be taught to move their feet and stand correctly when the exhibitor pulls forward or pushes backward on the lead halter.

The beginner may be unable to remember everything or always have the feet in the right position as the judge walks around the ring. For the younger exhibitors it's more important to have the animal standing in an attractive position, rather than to be constantly moving rear feet. This is where training before the show is important. If the animal is hard to pose at home, it will likely be worse in the show ring.

Allow plenty of space between animals when stopping the heifer for the judge, or placing her in a line. The judge can see your heifer better if she is not crowded against other animals. Allowing extra space to the front lets you pull up a few steps if the exhibitor behind gets too close.

You should continue showing while the judge gives placings and reasons. Everyone can learn from the judge's reasons regardless of where they are placed in line. Courtesy and sportsmanship are always important in the ring.

Watching some of the other classes being judged is a good way to learn. It's easier to understand what the judge is saying when you aren't directly involved with what's going on in the ring.

FEEDS & RATIONS

Ruminant Nutrition

Of all the movies we used to see in vo-ag class "Digestion in the Ruminant" must have been the best. It's the one I remember best, anyway. This movie was filmed from inside a cow's stomach.

I have no idea how they got the camera inside a cow, but it certainly was effective. Each time the cow would swallow, a wad of forage would come splashing down into the rumen or the reticulum or wherever the camera was focusing on this obscene mass of digestive material.

Everything was gurgling and sloshing like the inside of a washing machine. Viewers felt as if they were afloat in a boiling sea of gaseous digestion.

The narrator talked about antiperistaltic actions and other things most of us didn't care much about. It seemed that nutrition class always followed lunch, and after a few minutes of this movie students were wishing they had skipped the chicken à la king.

One reason for showing the film was to make ruminant nutrition exciting for vo-ag students, and it worked. The weaker your stomach, the more exciting it was.

It's hard to make nutrition exciting for 4-H and FFA members, but a peak inside a cows stomach helps explain a lot of things about feeding ruminant animals.

Ruminants have several compartments in their stomach and chew a "cud". Cattle, sheep, and deer are examples of ruminants. Other species, such as horses and rabbits, also digest forages very well — but have only one stomach compartment. Horses and rabbits perform much of their digestion of forages in the cecum.

It takes about four days for many common rations to pass through the digestive tract of cattle, as compared to the 24 hours required for passage of feed through a single-stomach animal, such as a pig.

Concentrate feeds, such as grains, are digested faster than forages. Bulky feeds, such as straw, poor quality hay, or low quality pasture, are digested slowly. While the cow or ewe may be very full of this low quality diet, she may be unable to digest it fast enough to meet her nutritional requirements.

Research has shown that a protein deficient diet tends to reduce appetite. The animal on a low protein diet has a tendency to eat less, as well as being unable to digest the low quality forage fast enough.

An empty digestive tract stimulates appetite in ruminants. Animals fed slowly-digested forages will have less appetite than those eating easily digested feeds, such as grains.

Because cattle chew their feed only partially, the digestion time and efficiency of digestion can often be speeded up by feed processing. Lower quality forages are digested faster when pelleted. Feed efficiency for cattle is generally improved by rolling or pelleting grains.

Micro-organisms in the rumen are important to the digestion of roughages and enable the ruminant to utilize non-protein sources of nitrogen, such as urea. Because young calves and young lambs don't have a fully functional rumen they can't utilize non-protein nitrogen. Rations formulated with urea can be toxic to young animals.

Bacteria use nitrogen and other elements available in the rumen to construct amino acids for building their own bodies. When the rumen bacteria are digested by the ruminant animal, these amino acids are used for making protein.

The balance and type of micro-organisms in the rumen changes with different types of feeds. A change in feed gradu-

Ruminants digest forages very well, thanks to their four stomach compartments and micro-organisms in the rumen.

ally alters the type of micro-organisms present in the rumen. A sudden ration change often causes digestive problems.

Most absorption of protein and carbohydrates occurs in the first part of the small intestine. The term "bypass-protein" refers to proteins that escape fermentation in the rumen and are digested in the abomasum and small intestine; thereby retaining protein quality more effectively than proteins which are subjected to rumen fermentation. Feeds containing significant amounts of bypass proteins are corn gluten meal, bloodmeal, dehydrated alfalfa, and milk products.

So what does this mean if you are feeding steers, lambs, or breeding stock as 4-H and FFA projects? First, it means we had better read the feed tag and make sure the ration has been formulated for the type of animal we are feeding.

We don't want to give "cow feed" containing non-protein nitrogen to young lambs or young calves. This feed wouldn't provide the desired protein and may be toxic to young animals.

For top gains, we need to feed animals at a regular time

each day, and to change rations gradually rather than abruptly. If it takes four days for a typical ration to pass through the digestive tract, we should expect the animal that goes "off feed" to recover over a period of days, rather than hours.

Grains should generally be rolled, cracked, or pelleted for a cattle ration (with certain exceptions.) Lambs apparently digest whole grains more completely than cattle do and will do fine on rations with unprocessed grains.

Much of what goes on in the rumen requires water. Adequate water intake is essential for proper digestion.

Hay Quality

If our animals could talk, they would surely tell us a few things about the hay we feed them. The quality of our forage affects all other aspects of the feeding program for ruminants.

As a rule, there is no class of animal that really likes poor hay. We may feed them lower quality hay if it meets the animals' requirements and is more economical for that class of livestock, or if we want to get more fiber in the diet. Otherwise, good quality hay is preferred.

Most feeding information divides hays into categories, such as grass hays, legumes, or grass and legume mixes. In many areas of the country alfalfa is the most common legume hay, and other types are compared to alfalfa when discussing quality. Many rations will be based upon "good quality alfalfa."

How can we identify good quality alfalfa? First it should be green and leafy. Although a bright green color doesn't guarantee high quality, it generally indicates proper curing and a high carotene content. On the other hand, some bleaching and loss of color doesn't always indicate poor quality.

Maybe we should ask, "What does high quality alfalfa feel like?" Some experts say the feel of hay is at least as important as visual appraisal. The softness of hay is related to maturity at cutting, leaf content, and good curing methods. All are important factors in quality.

Two thirds of the protein in alfalfa hay is found in the

leaves. Any loss of leaves during curing and baling can result in major losses in quality.

Hay quality is best estimated by a laboratory analysis in combination with visual appraisal. Many hay growers will have chemical test results and can provide information on protein content and estimated digestibility for all of the hay they sell.

The single most important factor in hay quality is the stage of maturity at cutting. Protein content, palatability, and digestibility all decrease as the crop matures. Early cut hay is more desirable in all of these factors.

The main visual indicator of stage of cutting for alfalfa is the amount of bloom present (if any). The number and size of seed heads is an indicator of maturity in grass hays.

The traditional recommendation for the best balance of quality with quantity has been to cut alfalfa hay at 10 percent bloom. The best cutting stage varies with growing areas as well as the final use for the hay.

Higher protein isn't always better. Mixtures of grass and alfalfa are excellent hays for many classes of ruminants. Grass hay is often preferred over alfalfa when finishing 4-H or FFA steers to maintain adequate fiber in the diet and reduce digestive problems.

A major difference between grass hays and alfalfa is the protein content and a much lower incidence of bloat from grass forages than from alfalfa. This is sometimes important when feeding high energy rations to cattle.

Although early-cut grass hays can have a high protein content, many of these forages aren't cut that early. The protein level declines rapidly as the grass matures, and we often have to add extra protein to rations for young animals when grass hays are a major part of the ration.

Ruminants need a certain amount of roughage in the rumen to maintain proper fermentation. This is variable with age and type of animal as well as what type of grain or concentrate is being fed.

When finishing steers or lambs we generally limit the

Stage of growth at cutting is the single most important factor in hay quality.

roughage and increase the grain portion of the ration in order to get more energy into the ration and achieve a faster gain. In these situations it's common to limit the amount of hay fed to 20 percent of the total ration or less.

4-H and FFA members must remember that quantity and quality of hay fed has a big effect upon grain consumption. One of the most common mistakes in feeding project lambs or steers is to throw them a "flake of hay" twice a day and thereby wind up with too much forage in the ration. Hay is like snow: No two flakes are alike.

When lambs are being finished on pelleted rations, many feeders provide no long hay. The pellets contain forages, however, and have a fairly high fiber content. If straw is used for bedding we might notice the lambs are eating the straw and getting more roughage than we realize.

Management of Small Pastures

The late Ace Reid drew a cartoon many years ago that shows a thin cow against a landscape of dirt and rocks. One cowpoke is saying to the other, "That's why my cattle are heavier than yours. Mine eats rocks!"

In the quest for high-producing livestock we must give equal attention to the animal's nutritional requirements. For ruminants and horses this generally includes pasture management.

The part-time farmer or family with a few 4-H animals faces some unique pasture management decisions. Because profit isn't the main motive, most owners of small acreages have more animals than their pasture can support. This doesn't alleviate the need for management, but it changes the approach.

The best course of action for most small landowners is to make maximum use of pasture forages in the spring or early summer and then confine the animals to a smaller area, often called a "dry lot" for the rest of the year. This is safer for the animals and allows proper management for the remaining pasture.

When animals are confined to overgrazed pastures, they often eat everything except the poisonous weeds. Then, after a while, they go ahead and eat the poisonous plants, too. The smaller dry lot is preferred because all vegetation has been removed by the animals' hoof action and intense grazing.

It's better to have 1 acre of dry lot and 9 acres of good pasture than to have 10 acres of noxious weeds.

Pasture management requires grazing at the right time for the grasses and/or legumes present and having the animals off long enough for the plants to make regrowth. Regrowth must occur between grazings and between growing seasons for the plant's root system to rebuild, and for a pasture to remain productive.

Intensive grazing for a short period of time is generally better for the plants than continuous grazing throughout the growing season.

Getting maximum production from a pasture requires fertilization. We can double or triple the production of grass pastures with an application of nitrogen fertilizer. If the pasture is producing one ton of forage per acre (on a dry matter basis) and production can be increased to two tons, that's a ton of hay we won't have to buy. Of course irrigated pastures and those receiving plenty of rainfall will do much better than this.

The fertilizer required to increase pasture production is much less expensive than the hay this added production replaces. Possible exceptions might include areas where pasture leases or hay prices are very low.

The amount and type of fertilizer required will vary for different areas of the country and the forage species in the pasture. County extension agents and Natural Resources Conservation Service (formerly Soil Conservation Service) employees can provide recommendations to fit local conditions.

Nitrogen fertilizers can be detrimental to legumes in mixed pasture stands. Nitrogen benefits grasses much more than legumes, and this tends to increase the grasses at the expense of the legumes. In many instances the increased yield of grass from nitrogen application outweighs any detrimental effect on these mixed pastures.

The Natural Resources Conservation Service or county extension office is also a good place to find recommendations on seeding new pastures. These folks will suggest

Fertilization is necessary to achieve top production from grass pastures. In some areas irrigation is critical, also.

adapted forage species and help the landowner avoid problems sometimes incurred with pasture mixes.

Many commercial outlets sell pasture mixes which include a large number of grass species, on the theory that if you plant enough different species, something is bound to grow. If the buyer doesn't know something about the species desired, it's not uncommon to get a mix containing a lot of things you don't really want.

Aside from paying for grass species you don't want, a second problem with seeding several different species is that each grass has different growth habits and management requirements. If these aren't compatible, it's difficult to devise a grazing schedule to fit the species present.

Animals have no sympathy for grasses: They will eat the youngest most palatable grass until it is gone, before grazing the older more mature forage.

How can a backyard farmer estimate how many animals a pasture can support? For me the easiest way is to estimate the amount of dry forage one might get if this forage were baled.

If you can look at an area and say, "This pasture would

make about a ton of hay to the acre," you can estimate the amount of forage available for the animals. The old "take half, leave half" grazing philosophy says a pasture that would produce a ton of hay per acre will support grazing equal to 1,000 lb. of dry forage per acre.

If your cow normally eats 20 lb. of hay per day, an acre of this pasture would support one cow for 50 days, two cows for 25 days — or 10 adult sheep for about 20 days.

It's not quite that simple, of course. When and how quickly forage is grazed is important.

Fencing to allow rotation of pastures and shorter duration of grazing will increase the carrying capacity considerably. The "take half, leave half" philosophy is O.K. for a beginning estimate, however.

Feeding Wheat

Although wheat is not considered a feed grain in some parts of the country, there are situations where this grain becomes a viable feed alternative. Sometimes wheat is an economical feed because of a poor wheat market, or because of reduced quality from sprout damage or contamination with other grains.

Transportation and selling costs in the wheat growing regions sometimes make it more economical to feed wheat on the farm than to haul it to town and transport other grains back to the ranch. This has been true during poor wheat markets in the past. Maybe wheat won't get that cheap again. Who knows?

While some folks have learned how to feed wheat, others have not. There are a lot of scary stories about wheat being a "hot feed."

It pains me to recall how many times I've heard 4-H and FFA members told a particular feed was too hot, or would "burn-up" their animals. I figure you can make them sick, and you can kill them; but you won't burn them up.

Wheat can provide a significant portion of the grain ration for all types of livestock and can be the sole grain fed to some classes of animals. This grain may be fed as the only energy source for poultry and swine if properly supplemented.

Because it can cause rumen disorders if fed at high levels, wheat generally makes up less than 50 percent of the ration when fed to cattle and sheep. When wheat is fed as less than 50 percent of the grain ration, most studies show it to be worth about the same as shelled corn for feeding steers. Coarse grinding is recommended for wheat fed to cattle.

Wheat is a good energy source for market lambs or sheep breeding stock and grinding is not required for either lambs or breeding sheep, except for old ewes that may be short on teeth.

Although some studies have shown wheat can be fed as the total grain ration for market lambs, most experts recommend mixing with barley or oats so that wheat makes up less than 50 percent of the ration.

Because of the higher energy content of wheat and the potential for mistakes, this grain may not be quite as foolproof as something like barley, oats, or prepared pelleted feeds.

Junior livestock raisers must learn to weigh their feed, rather than measure it. This is especially true with a heavy grain, such as wheat.

A coffee can full of whole wheat weighs nearly twice as much as the same can full of whole oats. If the feeds are ground, the weight-per-can changes; and when you buy a new batch of grain, the weight may change again.

It's important to feed enough roughage when feeding wheat to lambs or cattle. It's not like the pelleted feeds that contain a percentage of hay.

Wheat is an excellent feed for swine and has performed even better than corn in many studies. Some trials have shown better feed efficiency from pigs fed wheat than those fed corn, although the rate of gain may be slightly less for those on the wheat ration.

The protein content of wheat is higher than corn; but when fed to swine, supplementation is required for added protein and for amino acids such as lysine. There is a great amount of variability in the protein content of different types of wheat, with the hard red varieties normally providing more protein than soft red or soft white varieties.

It is generally recommended that wheat be coarsely ground or pelleted for swine rations, but it can also be fed whole. If whole wheat is available at a much lower cost than coarsely ground or rolled wheat, a slight increase in feed efficiency from grinding may not offset the increased cost of this operation.

If a 4-H or FFA member has only a few pigs and ready access to the wheat in Dad's bin, this whole wheat may be preferable to the more expensive rolled wheat available in town. Some references suggest that whole wheat be self-fed to permit time for the pigs to chew it. They tend to gulp it down when hand fed. Many pig feeders soak wheat in water overnight on the theory that this improves digestion of whole kernels.

What about the sprouted wheat that sometimes results when poor weather at harvest causes the grain to sprout in the field? A University of Idaho study showed wheat containing up to 60 percent sprouted kernels to be nearly equal to sound wheat for swine rations. Feed efficiency was slightly less for pigs fed sprouted wheat, but their rate of gain was equal to those fed sound wheat.

Research at Washington State University showed no difference in performance or carcass characteristics for beef cattle fed wheat with up to 58 percent sprouted kernels, as compared to cattle fed sound wheat.

WHEAT RATIONS FOR FINISHING CATTLE

In the following rations, wheat makes up 40% of the total ration, when the roughage portion of the ration is also considered. These rations are formulated for roughage (hay) to be fed as slightly less than 15% of the total ration (example: 16 lb. grain with 2.4 lb. hay).

Ration 1:	%
Wheat	46.5
Barley	46.5
Molasses	5.0
Ground Limestone	1.5
Salt (trace mineralized)	0.5

To be fed with alfalfa hay

Ration 2:	%
Wheat	46.5
Barley	43.0
Molasses	5.0
Soybean Meal	3.5
Ground Limestone	2.0
Salt (trace mineralized)	0.5

To be fed with grass hay

Add vitamin A to both the rations at the level of 1000 IU per pound of feed.

Source: Washington State University Extension Bulletin EB1317 *Feeding Washington Wheat*

WHEAT RATIONS FOR FINISHING LAMBS

Ration 1:	%
Wheat	30.0
Barley	29.5
Salt (trace mineralized)	0.5
Alfalfa Hay	40.0

Ration 2:	%
Wheat	30.0
Barley	39.0
Ground Limestone	0.5
Salt (trace mineralized)	0.5
Alfalfa Hay	30.0

Ration 3:	%
Wheat	30.0
Barley	32.0
Soybean Meal	6.0
Ground Limestone	1.5
Salt (trace mineralized)	0.5
Grass Hay	30.0

Source: Washington State University Extension Bulletin EB1317 *Feeding Washington Wheat*

WHEAT RATIONS FOR PIGS

Ration 1: (Starter, 20-50 lbs., 18% protein)	%
Wheat	72.2
Soybean Oilmeal (47.5% protein)	25.2
Limestone	0.8
Dicalcium Phosphate	0.9
Vitamin Premix	0.5
Trace Mineral Premix	0.1
Salt	0.3

In the following rations, cull peas (a protein feed available in the Northwest) are substituted for a portion of the soybean oilmeal to lower the cost of the rations.

Ration 2: (Grower-Finisher, 80-130 lbs., 14% protein)	%
Wheat	77.6
Soybean Oilmeal	10.0
Cull Peas	10.0
Limestone	1.0
Dicalcium Phosphate	0.5
Vitamin Premix	0.5
Trace Mineral Premix	0.1
Salt	0.3

Ration 3: (Grower-Finisher, 130-220 lbs., 13% protein)	%
Wheat	79.5
Soybean Oilmeal	6.0
Cull Peas	12.0
Limestone	1.1
Dicalcium Phosphate	0.5
Vitamin Premix	0.5
Trace Mineral Premix	0.1
Salt	0.3

Ration 4: (Grower-Finisher, 130-220 lbs., wheat plus synthetic lysine)	%
Wheat	96.80
Lysine	0.35
Methionine	0.10
Limestone	1.20
Dicalcium Phosphate	0.70
Vitamin Premix	0.50
Trace Mineral Premix	0.10
Salt	0.25

Source: Washington State University Extension Bulletin EB1317 *Feeding Washington Wheat*

JUNIOR SHOWS

Winning and Losing

Those who hang around fairgrounds each summer are familiar with the cry of the discouraged parent: "This is the last year we are going to do this!"

I don't know why we keep saying that. I guess it relieves a little pressure. Heaven knows there is plenty of pressure at junior livestock shows.

Along with the fun and excitement of livestock shows, there are tensions for both parents and kids. Parental interest is always a good thing for junior livestock exhibitors; but there's a fine line between minor assistance and excess involvement.

Parents can make competition a positive experience for young exhibitors by helping them understand that winning isn't everything. If the only goal is to win, nearly everyone is going to be disappointed.

Winning and losing is important for everyone. People who lose all of the time don't adjust well to life's demands. Psychologists say those who win all of the time don't adjust well, either. Experience at both winning and losing is important for each individual.

Each of us has the almost daily opportunity to win or lose at something; but only on special occasions do we have the chance to do this in front of a crowd. For many youngsters, sporting events and livestock shows provide this opportunity.

I think attitudes toward competition are an individual thing. Through some combination of experience, social environment, and maybe heredity, each person decides what constitutes victory or defeat.

Most readers have seen the youngster, with the animal at the bottom of the class, break into a big smile when presented a white ribbon. We may have expected this young person to be disappointed at receiving what many consider to be a third-rate prize. But he or she felt good about their efforts and the ribbon was important.

Sometimes we may put too much emphasis on what color ribbon the kids receive, tending to emphasize the result rather than the effort. As a result, exhibitors may be happy with a poor job that nets a blue ribbon and disappointed with a good job that produces a red. The judge may compensate by awarding all blues to avoid too much disappointment.

We have all seen large numbers of young exhibitors upset when the judge gave only a few blue ribbons, and other classes pleased when the judge gave blue ribbons to the entire group.

It may be hard for the kids in the first instance to understand that a red ribbon may indicate good effort, or those in the second case to realize that the blue ribbon may not indicate exceptional performance. You will notice the second group is much happier, though.

The kids develop certain expectations, and as they grow older we see all sorts of reactions to winning and losing. Unfortunately, we see all sorts of reactions from their parents, too.

Some of these responses lead many adults to argue that competition is bad for kids; while others staunchly maintain that all of life is a competition. The most fervent competitors believe when the world comes to an end there will be one who survives; and that one shall be known as "the winner."

I'm not saying youngsters should be satisfied with less than their best effort, but it's important to keep things in perspective. We should emphasize the effort rather than the result.

This young lady has fun showing cattle. That's the way it should be.

The problem of cheating at junior livestock shows has received a lot of attention in recent years; and with good reason. Show ring cheating has become so severe in some parts of the country that people are going to jail for it.

Parental involvement is generally positive, but parents can get caught up in the competition of junior shows and perform stunts that kids wouldn't even think about. The problem of show ring cheating is almost always traced to parents who are obsessed with winning.

Cheating isn't just a question of ignorance or misunderstanding of the rules: It's a people problem. There have always been people who will cheat when they can, and then look you right in the face and say they didn't do it.

The best prevention for such illegal activities is intolerance. When other parents and exhibitors refuse to tolerate the unethical behavior of a few, these problems can generally be eliminated.

I witnessed a number of such incidents during my years as a county extension agent. Most of them were "nipped in the bud" because another parent or exhibitor reported the impending violation of the rules.

It's pretty hard to unload an illegal steer from a neighboring county fair, or switch animals two weeks before the show, without someone else in the club or chapter knowing about it. The rule-breakers generally straighten-up when they learn such behavior won't be tolerated by their peers.

People who bend the rules might get by with it sometimes; but they really haven't gained anything. Even if they make a lot of money from cheating, it isn't worth it.

Those who value their children will understand what I mean by that.

Buying Market Project Animals

One of the most important decisions for 4-H and FFA members is where and how to purchase a project animal. Starting the project with animals of a desirable breed, size, and age will go a long way toward assuring a successful experience. Of all the things that can go wrong in a livestock project, a bad decision at buying time is often the most damaging.

Selection of a project animal begins with finding a source of supply. Once this is accomplished, junior exhibitors should learn that paying a reasonable price and making your own selection is part of the learning experience. Not everyone will make money on their livestock project, but paying exorbitant prices in the hope of having the Grand Champion is not what these programs are about.

Finding the source of supply often requires some prospecting well ahead of buying time. This means talking with breeders to see if they will have animals of the right size and age for your show.

It pays to have more than one source lined up — in case breeders don't have the selection they expected. While steer exhibitors sometimes purchase animals 5 to 10 months before the show, youngsters showing lambs or pigs will have a narrow window of time to find a project of suitable size and quality.

Discuss the size and type of animal you want with your 4-H leader or vo-ag teacher before going to a farm to select a

Starting with a project of the right size and age helps assure a successful experience.

project. Don't expect the breeder to tell you what size animal you need (unless this breeder has recent experience with junior shows and a knowledge of your feeding situation).

Decide approximately how much you are willing to pay — before you go to the farm. Ask about price ahead of time to avoid surprises (and a lot of mumbling) when the breeder tells you what the animals will cost.

Despite my rantings about exorbitant prices, I do believe junior show animals are worth more than market price. We are asking the breeder to sell one of his best animals, and a better than average price is in order.

On the other hand I'm against paying double the market price on the assumption that money can buy the grand champion. The project members who pay reasonable prices generally do just as well in the show ring, and they certainly learn more about the economics of livestock production.

Junior exhibitors should be considerate of the breeder when selecting animals. Schedule a time that is convenient for him or her and show up on time.

If several club members are looking for projects from the same breeder, try to select these animals the same day. Livestock producers don't have time to pen animals for one buyer at a time on six different days.

While it is important for youngsters to select their own animals, I lean toward protecting younger members from certain disaster. If a youngster selects a severely under-sized or over-sized animal for a particular show, this can eliminate much of the learning that is supposed to take place.

A young person learns little about livestock feeding, for example, if the animal is already market weight when it's purchased. By the same token, an underweight animal may have no chance to make show weight — causing unnecessary discouragement and financial loss.

Some of these problems can be avoided if the breeder will sort animals of the right size and age, and pen these for the junior exhibitors to select from. This allows the youngsters to select their own projects with less risk of catastrophe. (It also keeps them from playing with the dog or swinging on the hay rope, when they are supposed to be looking at the animals.)

With that said, we might remember that the parents' objectives often don't match what the kids want to accomplish. If the project member wants the cute little pig with the flat nose — and it's about the right size — why not?

We should remember that most of us old guys have made a few bad guesses, too.

Reading Carcass Data

Carcass data programs are an important educational activity for junior livestock shows. Carcass information isn't of much value, however, unless junior exhibitors and parents take the time to review the data provided.

A good way to study carcass data is to visualize the animals as they were when alive, and to compare this vision with carcass information. Your county agent, vo-ag teacher, or an experienced livestock producer can help explain various categories of data as they pertain to individual animals.

It doesn't take much of an expert to see that the amount and deposition of fat is the major element in determining meat quality and carcass value. Knowing how and why fat is deposited in various species of livestock can help us understand why one animal meets the desired grade while another doesn't.

Junior livestock exhibitors face the challenge of feeding animals to a desired market weight and a correct amount of finish for a particular show day. Recognizing the tendencies for fat deposition helps the exhibitor design a feeding program to accomplish this difficult feat.

Meat scientists tell us that young growing animals will use energy and protein for body maintenance until these nutrient needs are met. Only when we provide more nutrients than needed for muscle growth and maintenance, will the animal deposit fat.

The first fat deposited is within the body cavity. This is

kidney, heart, and pelvic fat. The second priority for fat deposition is subcutaneous, such as over the back, ribs, etc.

After these priorities are met the animal begins to deposit fat within the muscle. This intramuscular fat is called marbling in steers and feathering or streaking in lambs and hogs. A certain amount of fat within the muscle is considered important for desirable meat flavor.

Of course, some fat is being deposited externally and within the muscle at the same time the kidney, heart, and pelvic fat is forming. The priority is for internal fat, followed by subcutaneous fat and then intramuscular fat, however.

Problems arise when we expect steers to have enough fat within the ribeye muscle to meet the choice grade but very little external fat. This requires a compromise. Show judges and meat packers have to accept a certain amount of external fat in order to have steers with enough marbling to meet desired quality standards.

The process of fat deposition is reversed when the animal begins to lose weight — because of a reduction in feeding or an increase in stress. In steers this means the intramuscular fat (marbling) is the first to go when the animal is losing weight or condition.

That's why it is suggested that steers should be on a gaining plane of nutrition the last few weeks before show. If the marbling is lost due to reduced nutrition or increased stress, the animal's carcass won't make the choice grade.

In contrast, meat quality is much less affected (if at all) when lambs or pigs are held at a reduced gain, unless this dieting is carried to the extreme. This is partly because lambs and pigs need much less intramuscular fat for desired meat flavor and quality than steers do.

Nearly all lambs shown in junior livestock shows have enough fat streaking to meet the choice grade, and nearly all pigs in these shows have enough intramuscular fat to qualify for the number "1" grade. So we don't have to worry much about producing enough fat streaking in lambs or pigs fed for shows. We are more concerned about keeping the external fat (backfat) to a low percentage of the total carcass.

The number "1" grade for pigs requires less than .67-inch of backfat per 100 lb. of carcass. This grade is also dependent upon quality factors, such as firmness and color of lean and fat, and fat feathering between the ribs. Once these criteria are met, a combination of the backfat measurement and muscle score determines carcass grade for hogs. The less backfat the better.

Choice and prime grades in lambs are based upon similar meat quality factors as well as needing a minimum amount of fat cover and muscling to qualify as a choice carcass. It's not hard for a lamb to make the choice USDA grade. Of the 80 to 100 lambs shown at our local fair each year, I can't remember seeing one that didn't grade choice or better.

Most lamb carcass programs suggest lambs have a minimum of one-tenth inch of backfat. That's not much. When you feel a live lamb with this amount of fat cover, you can easily feel the vertebrae down the back. For pigs we want less than .67 inch average backfat per 100 lb. of carcass.

When you get your copy of the carcass data this year, take a look at the columns containing the external fat and internal fat measurements for your animals. Consider the animal's body type, age, and the feeding program.

Then look at these figures for your friends' animals. You will find that carcass data isn't so hard to understand after all.

Hiring Livestock Judges

The first rule for hiring livestock judges is "There are no rules for hiring livestock judges." The second rule is "Start early."

Finding qualified junior livestock judges and letting them know what your show expects takes time and communication. Whether the county agent or the show superintendent accepts this responsibility, many 4-H leaders and other volunteers will be faced with this task sooner or later.

Even if your show counts on the county agent to hire judges, this person can do a better job when there is good communication with show superintendents.

The first step in hiring good judges is understanding the judge's goals and the problems he or she faces. This requires pre-planning and communication so that everyone knows what is expected.

The first contact with a judge should normally be six months or more before the show. This allows this person enough time to schedule around other commitments or arrange a couple of days off if needed. Another contact should be made a few weeks before the show to discuss details or to send information about show rules and requirements. A final letter a week or two before the show reminds the judge of the date, time, and location of the show.

I learned a long time ago that if it isn't in writing, nobody knows for sure what was decided. If the first contact with a judge is by telephone, some written communication should follow the verbal agreement.

It helps to send copies of these letters to others involved with the show. This keeps everyone informed and protects against mistakes, such as saying Monday, August 15 — when August 15 happens to be a Tuesday. Sending copies to others is a good way to cover your backside, as they say.

One purpose of the initial contact is to inform the judge of specific rules or standards for your show that may differ from other shows. Anything the show management feels strongly about should be discussed with the judge. If your show has some requirements the judge doesn't believe in, this person has a chance to decline the offer, rather than go against what he or she believes.

We shouldn't be afraid to talk about money, either. I didn't say judges' fees weren't important. I only suggested that other things are important, too. Being able to offer a reasonable fee, plus expenses to judges makes the job much easier for the person who is in charge of hiring them.

As one who used to hire a lot of judges, I have never seen a fee that was too high. Judges take a day or two away from the farm or business, give up a day's vacation, travel long distances, eat unusual foods (and dust) in order to judge the show. They deserve a decent fee.

How do you decide who to hire? Ideally you would hire someone you have seen judge at another show. If this isn't possible, the next best thing is to go on recommendations of others.

Some states print livestock judges lists to help those looking for a show judge. While these lists are better than they used to be; it's hard to keep a list up to date or to set a standard for judges' qualifications.

Looking back, I can see that one of the biggest mistakes made in hiring judges is not talking with them enough before the show. The tendency is to accept a recommendation and then hire the judge without much discussion of the show's requirements or the judge's beliefs. On the other side of the coin, the judge may prefer to decline the offer if the show has requirements he or she can't work with.

The need for good communication doesn't stop when the judge arrives at the show, either. Even the most experienced judge appreciates being reminded of show customs and requirements that may differ from one area to another. A well-organized show superintendent is a great help to the judge.

We shouldn't expect the judge to be the sifting committee for the show. This person should not be asked to determine if animals are purebred or if they are in the right class, for example. Classes should be made up by show management and judged accordingly.

By the same token, the judge should not be asked to interpret rules, such as, "Do you think we should disqualify these steers with all of this spray goop on them?" It's not up to the judge to enforce the rules.

Large classes should be divided into classes of a manageable size. Part of the communication before the show might be to tell the judge how many kids are involved in showmanship and ask what size classes are preferred. This is better than having to split classes after the showmen have already entered the ring.

Ideally, the show superintendent and judge would work as partners to make the show as educational as possible. Then when someone asks, "How would you like to judge the guinea pigs?" the swine judge is likely to be a good sport about the whole thing.

Junior Livestock Sales

Few events have been the subject of more applause, criticism, and general debate than the junior livestock sale. Sometimes it's difficult to keep the sale in its proper perspective.

Although I have no particular position or authority to explain junior livestock sales to others, I do have some opinions. All of my opinions come with a disclaimer: They may be changed or altered at any time as a result of aging, thinking, or being beaten into submission.

As far as I can tell, the purpose for the first junior livestock sales was to get rid of the animals. I have been told this was necessary to prevent diseases being taken back to the farm with market livestock from the show. That doesn't make total sense to me, but it sounds O.K. for now.

I believe the second purpose of the sale is to reward the kids for their work and to give them some money. Buyers give the kids money to help them pay project expenses, save money for college, or buy a new motor-scooter. Different buyers have different reasons.

I was talking to a friend some years ago when he told me about four sheep he had bought. "What are you doing with four sheep?" I asked.

"I bought them from the grandkids," he said. "Whenever one of the kids needs some money, I buy a cow or a sheep from them."

It's nice to have a way to give the kids some money

occasionally. That's why we have scholarships.

A third purpose of the sale is to support 4-H and FFA programs. This may be somewhat confusing, because little if any of the money spent at the sale goes to the total 4-H or FFA program. It all goes to the kids.

People who buy the animals want to support the kids and their projects, though. And this is their way of supporting the junior livestock program.

I'm sure there are other reasons to have a junior sale, but that's a start.

Now, how can the junior sale be improved? I believe we need to keep our purposes in mind if we want to make improvements.

If furnishing animals of the desired weight to the meat packer is not a major purpose of the sale, I don't get excited about changing sale rules to get two more cents from the packer.

I'm not saying that to be nasty, but to point out that while there are reasons for all of the things we do in an attempt to improve a show or sale, it's real easy to lose sight of our goals when we start making rules.

If selling rabbits through the junior livestock sale meets all of the sales purposes, why not sell the rabbits (as many sales do)?

If a major reason for the sale is to give the kids some money, we should encourage buyers and make them feel good about the investment they are making.

"Thank you" letters from the kids are very important, and many clubs or sale committees furnish pictures of exhibitors and their animals, with certificates for buyer recognition. These activities not only encourage buyers for the sale, but also generate good feelings and support for youth programs.

When I was young county agent many years ago, I had thoughts like other young agents. I sometimes thought the livestock sale drew so much attention and time that it detracted from the total 4-H program. (Many people feel this way about the junior livestock program, or possibly the horse program.)

Then, I began to notice over the years that no matter how much we ignore something that is successful, we don't seem to make any faster progress with the things that aren't successful.

Then, I began to realize that things like junior livestock sales, and horse programs, and style shows help generate enthusiasm and support for other activities, too.

It's like the old saying, "The tide raises all boats." Meaning, we're all going up together, or we're all going down together.

So now, after many years of watching junior livestock shows, I tend to think, "If it ain't broke, don't fix it." The junior livestock sale is an excellent vehicle for increasing support for junior livestock programs as well as other youth programs.

Careers in Agriculture

It's the same old question: "What do you want to be when you grow up?" They start asking when you are a freshman in high school, and it seems they never let up.

Sometimes I just want to say, "How should I know? I'm only 50 years old!"

I'm sure younger people have the same problem. Everyone wants to know what you plan to do after high school, and sometimes that's a hard question to answer. The jobs are out there, though.

If you are a 4-H or FFA member who raises and shows livestock, why not major in animal sciences or some other phase of agriculture and build on something you enjoy? There is a real shortage of young people with the education and experience needed for many of the good jobs in agriculture.

Agriculture colleges are quick to point out that you don't have to grow up on a farm to do well in nearly any field of agricultural sciences. Most students in agricultural colleges don't have a farm background.

If agricultural sciences and industries were dependent upon students with a farm background, they would be in serious trouble. There aren't enough farm kids to go around.

On the other hand, I believe practical experience in agriculture gives a student important advantages for many careers. Kids who grew up on a farm, or raised 4-H and FFA project animals, have about 10 years of practical education

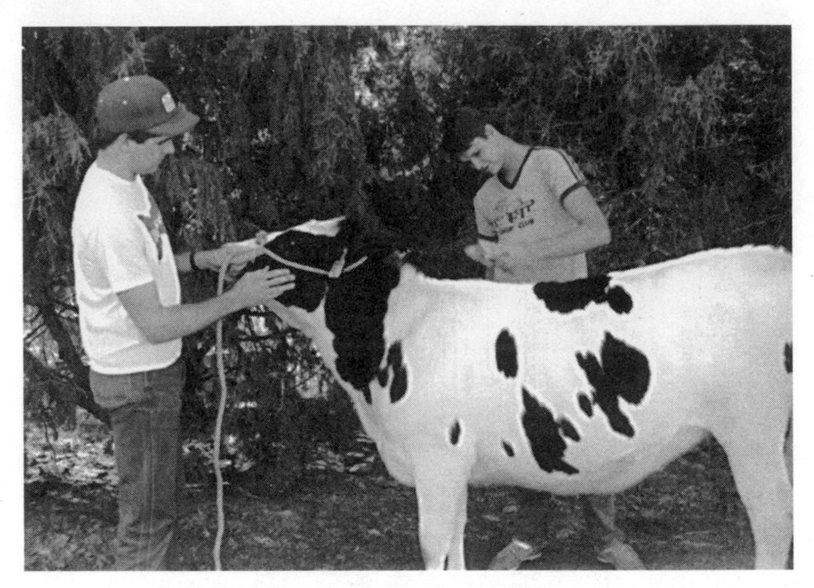

Practical experience with livestock can be part of a college education.

that can be very useful during and after college.

Let's take a look at some careers involving animals. Of course the most popular professional working with livestock is the veterinarian. Everybody likes veterinarians, and with good reason: They are intelligent, dedicated individuals — who are essential to the livestock industry.

There is no better career for those who have the ability and desire to become vets; but many students don't have that ability or dedication. That's why veterinary colleges are so selective: It saves some of us a lot of grief.

About half the students enrolled in Animal Sciences enter the university with plans to become a veterinarian. Most of them change their mind over time, or are unable to qualify for vet school.

That's O.K.. These are good students and welcome additions to agricultural colleges. They have a couple of years in college to decide if vet school is for them. If vet school doesn't work out, these students have a number of other opportunities as a result of their college degree and experience.

Who else do you know with a job in agriculture? The vo-ag teacher, certainly; and probably the county extension agent. The county agent also works with a number of specialists and other professors who teach and/or do research at the University.

Despite what we read about university budgets, there are lots of jobs in the university system. If you snoop around universities, you'll find all sorts of people in back rooms and basements, checking the pulse of an armadillo, or trying to learn if tire marks are hereditary in these creatures. Some of these scientists are also finding cures to important human diseases.

If you visit agricultural companies, you'll see they have researchers, too. Many of these people got started in their career because of an interest in animals.

What about the guy who runs the feed store? There's a good chance he has a college degree, and he buys feed from sales people who should know something about livestock if they're worth their salt.

The sales person may work for a distributor, who markets feed for a manufacturer that also hires researchers, nutritionists, sales people, etc. We are talking numerous jobs behind the products you see at the farm supply store.

All of these companies have marketing and sales departments, as well as many other jobs. I know high school and college students think they won't like sales jobs, but that's not being very realistic. Everyone is selling something — whether it's a product or an idea. (I've sold a few ideas that weren't nearly as good as most of the products on the market.)

There are jobs in the banking industry and the institutions that lend to farmers. These people certainly have to know something about what the farmer is trying to accomplish.

Then, there are farmers themselves. A high percentage of young farmers have a college degree, and you won't find many who regret getting that education.

What about agricultural journalism? All sorts of people

are writing about agriculture, and most of them don't know any more about agriculture than I know about journalism. There's plenty of room for people who know Polled Herefords are the ones without horns.

If I were advising high school students, I would encourage them to put their agricultural background and interests to good use. Contact your state college of agriculture. They'll tell you how to get involved.

APPENDIX

GLOSSARY

AMINO ACIDS - the basic constituents of protein.

ANTIBODY - various substances in the blood or developed in immunization which counteract toxins or bacterial poisons in the animal's system.

BARROW - castrated male swine.

BLACKLEG - a disease of cattle, often fatal, characterized by gangrenous swelling of the upper parts of the leg.

BLOAT - condition in ruminant animals caused by excess production of gases in the rumen.

BOAR - uncastrated male swine.

BO-SE - (pronounced Boe-C) injectable selenium used to prevent or correct selenium deficiency.

BRUCELLOSIS - any of a variety of infectious diseases caused by a parasitic bacteria, causing abortions in animals; also called Bang's disease.

BREEDER - often denotes someone who raises purebred stock; in the case of individual animals the breeder is the owner of the animal's dam.

BUMMER LAMB - orphan lamb or one that must be fed artificially.

BVD - abbreviation for 'BOVINE VIRUS DIARRHEA'; disease of cattle characterized by diarrhea, loss of appetite, and fever.

CASTRATE - removal of the testicles of the male animal.

COLOSTRUM - the first milk secreted by the animal, containing a large amount of protein and immunizing factors for the newborn.

COCCIDIOSIS - a disease in which the intestines are infested with a protozoan organism called coccidia.

CONFORMATION - the general shape of the animal as determined by its framework or skeleton and muscle structure.

CONCENTRATE - the high energy portion of a ration; usually made up of grains, or grains and a protein source, such as soybean oilmeal.

COMPLETE FEED - feed that can be fed alone; some pelleted feeds would be examples.

CREEP FEED - ration provided to very young animals; usually fed in location excluding access by larger animals by means of panels or other structures with small openings.

CROSSBRED - an animal whose sire and dam are of different breeds.

DAM - female parent.

DRESSING PERCENTAGE - the carcass weight divided by the live weight X 100.

DRYLOT - a relatively small area supporting little or no vegetation where animals can be confined.

ENTEROTOXEMIA - pulpy kidney or "overeating disease" caused by the sudden release of toxin by the bacteria Clostridium Perfringins Type D in the digestive tract of sheep; highly fatal; vaccine available.

ENZOOTIC ABORTION - in sheep characterized by premature birth and abortion.

EWE LAMB - ewe less than one year.

FEEDER PIG - pig that has been weaned and is ready for feeding; usually refers to pigs between 30 to 50 pounds; in junior livestock shows, may refer to any pig less than market weight.

FIBER - (in the diet) most fiber is made up of cellulose which can be broken down by the bacteria in the stomachs of ruminant animals (e.g., sheep, cattle) into usable carbohydrates.

FILL - the amount of food and water in the animal.

FLAKE OF HAY - what the kids tell you they fed the animals; something less than a bale.

FLUSHING - increasing the nutrient level or quality of the ration prior to breeding in an attempt to increase ovulation; a common practice in sheep and swine production.

FREE CHOICE - providing feed, water, or minerals in such a way the animals can eat or drink whenever they wish.

GESTATION - the period of pregnancy.

GILT - young female swine.

GRADE - animal that has one registered parent but is not eligible for registration.

GRAIN - seed of the cereal grasses, such as corn, barley, oats; often refers to the concentrate or high energy portion of a ration.

HAND FEEDING - feeding animals a specific amount on a regular schedule, such as twice a day.

HEAT CYCLE - period of time in which a female animal may be bred and conceive.

HEART GIRTH - circumference of the animal measured just behind the front legs.

HEIFER - young female bovine.

KETOSIS - pregnancy disease resulting generally from inadequate nutrition for the ewe the last three to four weeks of pregnancy.

LACTATION - period during which the female is producing milk.

LEGUME - plants of the family with fruit that splits on two seams, such as alfalfa, clover, peas.

MARBLING - fat deposition within the ribeye muscle; a factor in determining carcass grade for steers.

PACKER - person or company in the business of slaughtering live animals and selling meat wholesale.

PARTURITION - giving birth.

PEARSON'S SQUARE - method of calculating ratios of two ration ingredients in order to meet desired nutrient requirements.

PREGNANCY TOXEMIA - see "ketosis".

PROTEIN (crude) - all the nitrogen-containing compounds in a feed; (digestible) - approximate amount of protein that will be digested by the animal.

PROTEIN SUPPLEMENT - feed of high protein content which is added to increase the protein level of a ration.

PROLIFICACY - production of numerous offspring.

PUREBRED - animal which is registered or eligible to be registered.

RATION - everything available for ingestion by the animal; includes concentrates, hay, pasture, browse etc.

REGISTERED - animal which is recorded with a recognized breed association.

RIBEYE MUSCLE - muscle located along both sides of the backbone; used in determining quality grade and calculating cutability in beef slaughter animals.

ROLLED GRAIN - grain which has been processed with a roller mill.

RUMEN - the largest compartment of the ruminant animal's stomach, in which the major roughage utilization occurs.

SELF-FEEDING - making feed available to the animals at all times.

SHOT - an injection.

SHOWMANSHIP - ability to show; a class in junior livestock shows which judges the ability of the exhibitor to prepare and show the animal in the ring.

SIRE - male parent.

STEER - castrated male bovine.

TAGS - matted wool; separated from fleece and lamb's wool when shearing.

TDN - Total Digestible Nutrients; commonly used as a measure of energy in a livestock ration.

VACCINATION - introduction of antibodies into animals causing them to produce an immunity or tolerance to a disease.

VIBROSIS - in sheep and cattle characterized by abortion in late pregnancy or birth of dead or small fetuses.

WEANER PIG - term used in some areas to denote a recently weaned pig, usually 25 to 40 pounds.

WEANING - separation of the young from the dam.

WETHER - castrated male sheep.

WITHERS - area over the top of the shoulders in dairy cattle.

REGISTRY ASSOCIATIONS

Sheep

Alberta, Canada Sheep Breeders Association
Norine Whiting, Sec'y
Box 2157
Claresholm, Alberta, Canada TOL OTO

American Border Leicester Assoc.
Kris Savage, Sec'y
Rt. 4, Box 138
Taylorsville, NC 28681

American Black Sheep Registry
4714 Glade Rd
Loveland, CO 80538

North American Clun Forest Assoc.
W5855 Mahlun Rd
Holmen, WI 54636

American Corriedale Association
Russ Jackson, Sec'y/Treas.
2911 N. 32nd Rd
Seneca, IL 61360

California Red Sheep Registry Inc.
Janice Altamore, Registrar
1850 E Reilly Rd
Merced, CA 95340

Columbia Sheep Breeders Assoc.
Richard L. Gerber, Sec'y
PO Box 272W
Upper Sandusky, OH 43351

American Cotswold Record Association
Vicki Rigel, Sec'y
18 Elm St., PO Box 59
Plymton, MA 02367

N. American Dairy Sheep Association
Tanya Gendreau
N20712 Thompson Lane
Galesville, WI 54630

American Delaine-Merino Record Association
Elaine Clouser, Sec'y
1026 Co. Rd 1175 RD6
Ashland, OH 44805

American Dorper Sheep
Breeders Society
2427 220th St
Tripoli, IA 50676

Continental Dorset Club
Marion Meno, Sec'y
PO Box 506
Hudson, IA 50643

Finnsheep Breeders Association
Claire Carter, Sec'y/Treas.
Phone: 317-297-3670

American Hampshire Sheep Association
Karey Claghorn, Sec'y/Treas.
PO Box 277
Whiteland, IN 46184

Jacob Sheep Breeders Association
Jane Fenton, Sec'y
6350 E Co. Rd 56
Fort Collins, CO 80524

Jacob Sheep Conservancy Breed Registry and Organization
9241 Eureka Rd
Girard, PA 16417

American Karakul Sheep Registry
Julie De Vlieg
Rt 1 Box 179
Rice, WA 99167

Katadin Hair Sheep International
Laura Fortmeyer
PO Box 115
Fairview, KS 66425

National Lincoln Sheep Breeders Association
Roger Watkins, Sec'y
PO Box 277
Whiteland, IN 46184

Navajo-Churro Sheep Association
Box 94
Ojo Caliente, NM 87549

Montadale Sheep Breeders Assoc.
PO Box 603-H
Plainfield, IN 46168

Natural Colored Wool Growers Association
Carole Sanders, Registrar
PO Box 487
Willits, CA 95490

American North Country Cheviot Sheep Association
Theresa Barefoot, Sec'y
PO Box 265
Lula, GA 30554

OPP Concerned Sheep Breeders Society
Mary C. Jarvis, Sec'y
2862 S Peterson Rd
Poplar, WI 54864

American Oxford Sheep Association
Mary Blome, Sec'y
Rt 1
Stonington, IL 62567

American Polypay Sheep Association
609 S Central, Suite 9
Sidney, MT 59270

American Rambouillet Breeders Association
Joann Custer, Sec'y
2709 Sherwood Way
San Angelo, TX 76901

North American Romanov Sheep Association
Don Kirts, Sec'y
PO Box 1126
Pataskala, OH 43062

American Romney Breeders Association
John H. Landers Jr., Sec'y
29515 NE Weslinn Dr
Corvallis, Or 97333

Scottish Blackface Sheep Breeders Association
Rt 3, Box 94
Willow Springs, MO 65793

American Shropshire Registry Association
Dr. Dale Blackburn, DVM, Sec'y
PO Box 250
Hebron, IL 60034

American Southdown Breeders Association
Gary Jennings, Sec'y
HCR 13 Box 220
Fredonia, TX 76842

American Suffolk Sheep Society
Annette Benson, Sec'y
PO Box 256, 17 W Main
Newton, UT 84327

National Suffolk Sheep Association
Dale VanHeuvelen, Business Manager
3316 Ponderosa St
Columbia, MO 65201

U.S. Targee Sheep Association
Debra J. Mrozinski, Sec'y
PO Box 310
Kouts, IN 46347

North American Texel Sheep Association
Linda Gayle Smith, Sec'y
740 Lower Myrick Rd
Laurel, MS 39440

National Tunis Sheep Registry
Lyle Hotis, Promotions
Rt.1 Box 192
Gouverneur, NY 13642

Beef Cattle

American Gelbvieh Association
10900 Dover St
Broomfield, CO 80021

American International Charolais Association
PO Box 20207
Kansas City, MO 64195

American Simmental Association
1 Simmental Way
Bozeman, MT 59715

American Angus Association
Richard Spader, Executive VP
3201 Frederick Blvd
St Joseph, MO 64501

American Hereford Association
PO Box 14059
Kansas City, MO 64108

American Salers Association
5800 S Quebec St 220A
Englewood, CO 80111

Beefmaster Breeders United
Wendell E. Schronk, Executive VP
6800 Park Ten Blvd, Ste 290W
San Antonio, TX 78213

International Brangus Breeders
Nell Orth, Executive VP
5750 Epsilon
San Antonio, TX 78249

American Brahman
Breeders Association
Jim Reeves, Executive VP
1313 La Concha Ln
Houston, TX 77054

American Blonde D'Aquitaine Association
James Spawn, Executive Secretary
PO Box 12341
Kansas City, MO 64116

American Chianina Association
Terry Atchison, CEO
PO Box 890
Platte City, MO 64079

American Highland Cattle Assoc.
4701 Marion Street Ste 200
Denver, CO 80216

North American Limousin Foundation
Dr. John Edwards, Executive VP
PO Box 4467
Englewood, CO 80155

American Maine Anjou Association
John Boddicker, Executive VP
1600 Genesee St, Ste 760
Kansas City, MO 64102

American Murray Grey Association
Jim Spawn, Executive Director
PO Box 34590
Kansas City, MO 64102

American Pinzgauer Association
Peg Meents, Secretary
21555 State Hwy 698
Jenera, OK 45841

American Red Brangus Association
Elaine Monagham, Director
3995 E Hwy 290
Dripping Springs, TX 78620

American Shorthorn Association
Dr. Roger Hunsley, Executive Secretary
8288 Hascall St
Omaha, NE 68124

Braunvieh Association of America
Iola Dieschot, Executive Secretary
PO Box 6396
Lincoln, NE 58505

North American Corriente Assoc.
Jim Spawn, Executive Director
PO Box 12359
Kansas City, MO 64166

Piedmontese Assn. of the U.S.A.
Harold Mayland, Sec'y/Treas.
4701 Marion St
Denver, CO 80215

Red Angus Association of America
Dr. Richard Gilbert, Exec. Secretary
4201 N Interstate 35
Denton, TX 76207

Santa Gertrudis Breeders Int.
PO Box 1257
Kingsville, TX 78364

Texas Longhorn Breeders of America
Carol Diffey, Executive Director
PO Box 4430
Fort Worth, TX 76164

United Braford Breeders
Dr. Rodney L. Roberson, Director
422 E Main St
Nocagdoches, TX 75961

Dairy Cattle

American Guernsey Association
Neil Jensen, Executive Secretary
761 Slate Ridge Blvd, PO Box 666
Reynoldsburg, OH 43068

American Holstein-Fresian Assoc.
1 Holstein Place
Brattleboro, VT 05301

American Jersey Cattle Assoc.
Calvin Covington, Executive Secretary
6486 East Main St
Reynoldsburg, OH 43068

American Milking Shorthorn Society
Betsy Bierdek, Executive Secretary
PO Box 449
Beloit, WI 53512

Ayrshire Breeders Association
Robert Schrull, Executive Secretary
PO Box 1608
Brattleboro, VT 05301

Brown Swiss Cattle Breeders Association of the U.S.A.
Michael W. Young, Executive Secretary
PO Box 1038
Beloit, WI 53512

Swine

American Yorkshire Club
Darrel Anderson
1769 US 52 W, PO Box 2417
West Lafayette, IN 47609

United Duroc Swine Registry
Darrell Anderson
1769 US 52 W, PO Box 2397
West Lafayette, IN 47609

American Berkshire Association
Joe Brown
1769 US 52 W, PO Box 2346
West Lafayette, IN 47906

American Landrace Association
Tom Park Jr.
1796 US 52 W, PO Box 2340
West Lafayette, IN 47906

Hampshire Swine Registry
Rex Smith
1769 US 52 W, PO Box 2807
West Lafayette, IN 47906

Chester White Swine Records
Dan Parrish
6320 North Sheridan, PO Box 9758
Peoria, IL 61614

Poland China Record Association
Jack Wall
6320 N Sheridan, PO Box 9758
Peoria, IL 61614

National Spotted Swine Records Association
Dan Parrish
6320 North Sheridan, PO Box 9758
Peoria, IL 61614

Index

A

A frame houses 123
Alfalfa hay 146, 163
Alfalfa pastures 64
Amino acids 112, 124, 126
Artificial insemination 46
Automatic waterers 127

B

B vitamins 124, 126
Baby pig care 128
Baby powder 120
Bang's Vaccination 144
Beef heifers
  breeding 45
  feeding 44, 48
  fitting 27
  health 28, 41
  selecting 37
  showing 30
Beet pulp 153
Blanketing dairy cattle 150
Blankets for lambs 94
Bloat 12, 64, 86
Blocking stands. *See* Trimming stands
Breaking steers to lead 13
Breed Associations. *See* Registry associations
Breeding age
  beef heifers 44
  dairy heifers 144
  ewe lambs 62
  swine 122
Breeding sheep
  breeding 59
  feeding 62
  fitting 68
  selection 58
Breeding swine
  breeding 122
  feeding 124
  health 123
  housing 123
  selection 121
Brucellosis 39, 141, 144
Bummer lambs 77
Buying project animals 183
Bypass proteins 161

C

Calf hutches 142
Calf starters 146
Calving 38, 51
Carcass data 186
Careers in agriculture 195
Cervix 74, 75
Choice grade 6, 9, 35
Clipping. *See* Fitting
Coccidiosis 89, 147
Coccidiostat 147
Cold milk feeding 77, 79

Colostrum milk for
 dairy calves 143, 146
 lambs 78
Crossbred
 ewes 58
 sows 121
Crude protein 86, 112, 113
Cull fruits 126

D

Dairy Cow Unified Score Card 137
Dairy heifers
 feeding 145
 fitting 149
 housing 142
 management 141
 selection 135
 showing 153
 training 149
Dehorning 141, 143
Dicalcium phosphate 147
Dilation of cervix 53
Dry lot 166

E

Enterotoxemia 88

F

Fat streaking 187
Feeding
 beef heifers 44, 48
 breeding sheep 62
 breeding swine 124
 dairy heifers 145
 market lambs 85
 market pigs 111
 market steers 8
Feeding wheat 170
Fertilizer 167
Fitting
 beef heifers 27
 breeding sheep 68
 dairy heifers 149
 market lambs 90
 market pigs 116
 market steers 17, 18
 stands 94
Flushing ewes 64
Frame size 4–6

G

Garbage 126
Gestation period
 hogs 122
Glossary 201
Grain overload 86
Grains for lambs 87
Grass hays 164

H

Hair wax 24
Halters for
 dairy 149
 lambs 95
 steers 15
Hand-feeding 86
Hay quality 163
Heart girth measurements
 dairy heifers 138, 140, 145
 pigs 109, 115
 steer 11
Heat lamps 129
Hernia 88
Hip height measurement 5
Hoof trimming. *See* Fitting

I

Intramuscular fat 187
Iron deficiency 131

J

Judging steers 34

K

Ketosis 66

L

Ladino 65
Lambing 74
Livestock judges 189
Livestock sales 192

M

Management
  beef heifers 44
  breeding sheep 62
  breeding swine 121
  dairy heifers 141
Marbling 36, 187
Market steers. *See* Steer
Market weights
  lambs 82
  pigs 107
  steers 4
Meat quality 187
Micro-organisms 160
Milk replacers
  dairy calves 145
  lambs 60, 77, 79, 80
  pigs 130
Mineral oil 120
Minerals 125, 126
Moldy grains 125

N

Noxious weeds 167

O

Orphan lambs 77

P

Pasture for
  beef heifers 46
  dairy heifers 147
  sheep 64
Pasture management 166
Pasture mixes 168
Pelleted rations for lambs 87
Pelvic measurement 39, 51
Pigs. *See* Swine
Pork Producers Handbook 131
Posing dairy heifers 155
Pregnancy toxemia 66, 67
Production records
  beef cattle 38
  dairy cattle 139
  sheep 59
  swine 122
Protein supplement 125

R

Rate of gain
  beef heifers 44
  dairy heifers 145
  lambs 85
  pigs 114
  steers 10
Rations for
  dairy heifers 147
  lambs 87
  pigs 114
  steers 9, 10
Raw beans 126
Raw potatoes 126
Registry associations 206
Retained testicle 88
Retention of afterbirth 52
Ribeye muscle 6, 8, 187
Ring steward 33
Ringworm 28, 41
Rotation of pastures 169
Rumen 160
Ruminant Nutrition 159, 164

S

Saddle soap 24
Salt 126
Scotch comb 21, 26, 30
Scours 87
Seeding pastures 167
Selecting
  beef heifers 37
  breeding sheep 58
  breeding swing 121
  dairy heifers 135
  market lambs 81
  market pigs 107
  market steers 3
Selenium deficiency 66
Self-feeding
  pigs 113
Shearing 68, 91, 101
Sheep (ewes & market lambs)
  breeding 58
  feeding 85
  fitting 90
  keds 71

meat breeds 70
selection 81
showing 95
wheat ration 170
Sheep dip 93
Show cane 116
Show stick 16, 30
Show whip 116
Showing
beef heifers 30
dairy heifers 153
market lambs 95
market pigs 116
Slick shearing 90
Soremouth 89
Steamed bonemeal 147
Steers
breaking 13
feeding 8
fitting 17
heart girth measurement 11
judging 34
rations 9
selection 3
wheat rations 170
Straw bedding 129
Swine (market pigs)
feeding 111
fitting 116
growth rate 108
health 111
selection 107
showing 116
wheat rations 170

T

Tail switch 22, 29
Tamed iodine 43
Tattoo 141
Trace minerals 146
Training
dairy heifers 149
market lambs 95
market pigs 116
market steers 17
Trimming stands 94, 95

U

Urea 126
USDA grade 188
Uterine bolus 76
Uterine contractions 52

V

Vitamin A 43, 50

W

Warts 28, 41
Washing. *See* Fitting
Water bag 52, 75
Weaning
dairy calves 143
lambs 79
pigs 122, 131
Wheat rations for
beef cattle 172
lambs 173
swine 174
White clover 65
Wool bags 102
Wool breeds 68, 70
Wool care 101
Wool contamination 101
Wool grades 104
Worming
pigs 111

Order Form

Pine Forest Publishing
314 Pine Forest Road
Goldendale, WA 98620
Phone: 509-773-4718

Please send______copies of
LIVESTOCK SHOWMAN'S HANDBOOK to:

Name ____

Address ____

City ______________________State ________ Zip ____________

Phone ____

Find enclosed a check or money order for ____ books, at $17.95 per book plus $2.00 postage and handling each.* Payment must accompany order. Make checks payable to *Pine Forest Publishing*.

Washington residents: Please add 7% sales tax ($1.26)

*Call or write for quantity discounts (5 or more).

✂ . . . .

Order Form

Pine Forest Publishing
314 Pine Forest Road
Goldendale, WA 98620
Phone: 509-773-4718

Please send______copies of
LIVESTOCK SHOWMAN'S HANDBOOK to:

Name ____

Address ____

City ______________________State ________ Zip ____________

Phone ____

Find enclosed a check or money order for ____ books, at $17.95 per book plus $2.00 postage and handling each.* Payment must accompany order. Make checks payable to *Pine Forest Publishing*.

Washington residents: Please add 7% sales tax ($1.26)

*Call or write for quantity discounts (5 or more).